INTRODUCTION TO
BRAIN TOPOGRAPHY

INTRODUCTION TO BRAIN TOPOGRAPHY

Peter K. H. Wong

University of British Columbia
and British Columbia Children's Hospital
Vancouver, British Columbia, Canada

With contributions by
Hal Weinberg and Roberto Bencivenga

SPRINGER SCIENCE+BUSINESS MEDIA, LLC

Library of Congress Cataloging-in-Publication Data

Wong, Peter K.H.
    Introduction to brain topography / Peter K.H. Wong with
contributions by Hal Weinberg and Roberto Bencivenga.
        p.   cm.
    Includes bibliographical references.
    Includes index.
    ISBN 978-1-4613-6653-9   ISBN 978-1-4615-3716-8 (eBook)
    DOI 10.1007/978-1-4615-3716-8
    1. Brain mapping. 2. Magnetoencephalography.   I. Weinberg,
Harold.   II. Bencivenga, Roberto.   III. Title.
    [DNLM: 1. Brain Mapping. 2. Magnetoencephalography.   WL 335
W872i]
RC386.6.B7W66   1990
612.8'22--dc20
DNLM/DLC
for Library of Congress   90-14228
                                                               CIP

ISBN 978-1-4613-6653-9

© 1991 Springer Science+Business Media New York
Originally published by Plenum Press in 1991

*This book is dedicated to
my mother, Wai-Yuk Kwan*

This book is dedicated to
my mother, Mrs. Nak Kaur

PREFACE

It had been difficult to find appropriate teaching material for students and newcomers to this field of brain electromagnetic topography. In part, this is due to the many disciplines involved, requiring some knowledge of the physical sciences, mathematics, neurophysiology and anatomy. It is my hope that this book will be found suitable for introducing interested workers to this exciting field. Advanced topics will not be covered, as there are many excellent texts available.

Peter K.H. Wong

ACKNOWLEDGEMENT

My co-authors, Hal Weinberg and Roberto Bencivenga, for their support; Richard Harner, for all his early advice; Ernst Rodin and Gene Ramsay, for their encouragement; Wendy Cummings for her assistance; Technologists from the Department of Diagnostic Neurophysiology for collecting such excellent data; Bio-Logic Systems Corp. for permission to use some data as illustration; and all my friends and colleagues.

My wife Elke, for putting up with me throughout this presumptuous endeavour.

The manuscript was delivered in camera-ready form to the Publisher. Illustrations were created using Harvard Graphics and CorelDraw software.

CONTENTS

FIGURE LIST

PART 1

FUNDAMENTALS

1.1 INTRODUCTION

This book provides an introduction to topographic EEG analysis. It is aimed at the practising neurophysiologist seeking an entry-level text, EEG technologists and EEG students seeking to bolster their knowledge, and the graduate student requiring help in their coursework. Material is presented and discussed at the novice and intermediate levels. Although fundamental concepts and some mathematical knowledge is covered, the emphasis is on the practical use of this new and exciting technique.

The rationale for topography is that the traditional EEG or evoked potential (EP) tracings contain information which under normal circumstances, is not appreciated by the naked eye. There is simply too much data, in a form unsuited for visual analysis. Frequency content of background EEG is one example. In this instance the solution is to apply mathematical processing by Fourier analysis, resulting in the familiar spectral plots. To discern the interrelationships between different scalp locations, one can merely arrange such plots in a manner as to mimic a head diagram, as had been demonstrated using 16 channels of EEG (Bickford 1977, 1981).

However, there are elements of the dataset which are still unrevealed: the exact location of the alpha peak, subtle asymmetries of the alpha distribution, complex phase relationships of waveforms etc. Although part of this information can be gleaned from the numerical representation, it is a laborious and non-intuitive process. It would be like trying to provide information of a city subway system by describing it entirely with words. A graphical communication of the same information (i.e., a system "map") would be more useful (Lehmann 1987).

Further argument can be applied to small time differences between channels (phase relationships). Two different epileptic foci may generate spikes that appear identical to the naked eye, due to the limited time resolution of ordinary EEG pen and paper write-out. Electronic displays with higher resolution would clearly reveal their intricate timing relationships. Coupled with suitable processing and display capabilities, a spatial-temporal analyzer for such EEG events can be implemented.

Quite apart from the display aspects of topography, there is a great selection of mathematical and statistical tools available to perform many different types of quantitative analyses. Such "post-processing" techniques cover a wide spectrum indeed. Examples include simple montage/reference reformatting and digital filtering; time-series studies, factor analysis, dynamical analysis; more advanced methods like source generator modelling/dipole localization and spatial-temporal modelling. Many of these are capable of shedding light onto hitherto hidden aspects of the data, or provide special insight by allowing interactive exploration of the datasets. It should be stated that the reader is not expected nor required to understand the intricacies of all the above techniques. They are covered within this text for the sake of completeness, and to allow the adventurous reader to wander further afield when the basics have been mastered. Indeed, it will be in one or more of these "esoteric" areas that one may find a potential solution to a current problem, or an approach that may portend a new and productive research initiative.

It is useful to view topographic analysis as a novel approach to clinical neurophysiology, to complement rather than replace the many time-proven visual analytical techniques. There are additional conceptual and technical knowledge requirements before one can skilfully extract the additional information which lies beneath the surface. To carry out clinical interpretation without adequate preparation would be inappropriate and fraught with difficulties. Likewise, to discard a new technique out of hand is equally inappropriate.

Topographic methodology includes the additional concepts of digital signal processing, computer graphics and cartography, numerical and statistical analysis, physics of electric fields, quantitative EEG and magnetoencephalography (MEG). These complex and diverse components are separated and covered in this and the following parts. Chapters within each part are designated for specific topics, and are thus relatively short. Throughout the book, there is a noticeable profusion of illustrations. This will hopefully make an otherwise difficult topic more palatable.

For the beginner, a small selection of normal data (VEP with flash and pattern, P300, EEG with eyes open and close) for different age ranges is presented in Part 5, as a guide to some normal variations. Space limitation does not allow a full display of this data, and thus only selected maps are presented.

1.2 DATA ACQUISITION

Using an array of recording electrodes, the scalp EEG signal is amplified and filtered. It then undergoes transformation from that analog form present at the amplifier output, into information suitable for computer assimilation. This step is the analog to digital conversion (ADC). The original EEG information is now in a digital form suitable for storage in computer memory, and ready for further digital processing and display.

There are 2 important parameters which govern ADC conversion: sampling rate (Hz) and precision (bits). These will be treated separately.

Fig. 1 - 1 depicts a single channel of EEG, as present at the output of the amplifier after suitable filtering. Two sampling rates were used. The vertical lines on the right denote the times when ADC occurs. By comparing this rate to the number of perturbations present in the signal in a given time window (say 1 sec.), one can appreciate that the first rate (top right) is low, and the digitized result is a crude representation of the original signal. The higher sampling rate (bottom right) produced a much smoother output, which was able to capture all the perturbations, or follow the changes of the signal faithfully. This "high fidelity" characteristic of digitization is important because it dictates the amount of temporal precision which can be achieved.

Inadequate sampling rate can occur if the frequency content of the signal is too high in relation to the sampling rate. Theoretically,

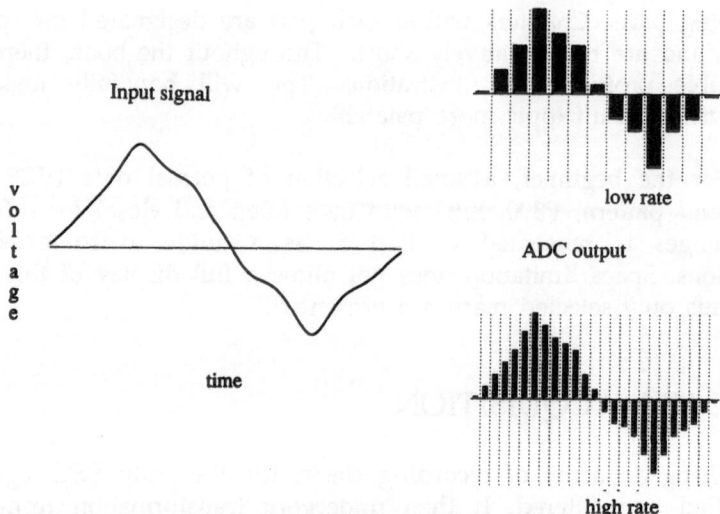

1-1:   Signal digitization. Left: input signal; right: output signals for different sampling rates (top=low rate, bottom=high rate).

the minimum sampling rate should be twice the highest frequency component contained in the signal. In practice, as the low-pass filters generally used during EEG amplification have a gradual roll-off in the frequency response curve, it is safer to sample at a minimum of 3 or 4 times the cut-off setting for the low-pass filter.

To illustrate the problems with inadequate sampling rate, Fig. 1 - 2 shows an input signal at one frequency being digitized into an output which contains some slower activity, which is obviously an artifact created by the sampling process. Such "alias" error is insidious and permanent, because it cannot be detected after ADC is completed, nor can it be removed. Along the same lines, an EEG input with high frequency beta may then possibly become contaminated with delta activity.

ADC precision is expressed as the number of "bits" which is used to represent any given voltage level. As Table 1 indicates, a precision of 8 bits means that there can be 256 such levels, while 10 bits means 1024 levels are available. Tied closely to this are the related parameters of dynamic range (through which the signal can vary without "clipping" or saturation distortions), and the minimum

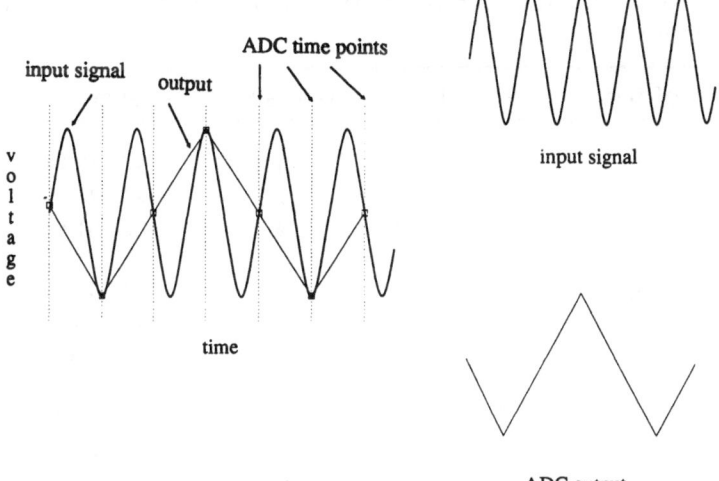

1-2: Sampling alias. The input signal is shown at the top right and on the left composite tracing. The dotted vertical lines are the times when ADC at a slow rate occurred, with the output a seemingly triangular waveform.

ADC voltage accuracy (ADC step size or "error"). If 8 bits were used to represent a range of 512 uV (or +/- 256 uV), then the absolute ADC voltage error is 2 uV, as any variation can only be in steps of 2 uV. However if a range of +/- 1024 uV is desired, then for the same ADC precision, the step size is 8 uV. If the requirements for the recording is such that this step size is unacceptably large, then the only alternative would be to increase the number of bits used.

Fig. 1 - 3 shows the effect of sampling at different precision. Here the input tracing is being represented by 6 bits, and shows the tracings obtained with decreasing precision.   At the minimum precision of 1 bit (bottom tracing) there are only 2 voltage levels: zero or maximum! However, the poor quality of the tracings may be hidden by the simple act of filtering, as may be done by software. As the right side of Fig. 1 - 3 shows, the smoothed tracings merely look better, but do not approximate more the input signal any better.

To put things in perspective, EEG spikes are of the order of 100 uV, while evoked potential waves can be from a fraction of a microvolt to 30 or 40 uV. In practice, it is useful to know that the ADC precision is usually determined by the electronic hardware

Table 1: Relationship between ADC precision and digitized data results.

# of bits	# levels	step size	range
8	256	1 uV	+/- 128 uV
		2	+/- 256
		4	+/- 512
10	1024	1	+/- 512
		2	+/- 1024
		4	+/- 2048
12	4096	1	+/- 2048
		2	+/- 4096

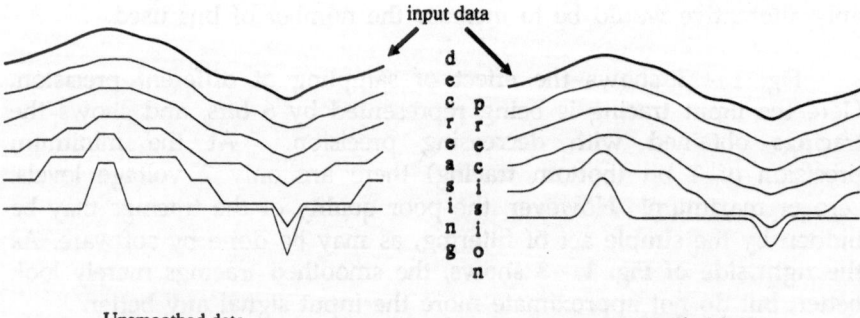

input data

decreasing

precision

Unsmoothed data   smoothed data

1-3: ADC precision. The input data is shown in bold line at the top, with decreasing levels of sampling precision (from 6 bits at input, to 1 bit for the lowest tracing). The visual effect of smoothing of the same data (right) is that of an artificially higher precision.

selected, so that the user has no choice. Based on this pre-determined precision, the user must make a compromise between step size and dynamic range. Again, the easiest way is to think in terms of the electrical short-circuit noise of the amplifier system. It is not useful to have the ADC minimum voltage step size much smaller than this noise value. So one selects the step size at the value closest to the amplifier noise. From this point on, the dynamic range is effectively determined.

Difficulty arises if this dynamic range is too small for the signal amplitude range. Then one must either accept a coarser ADC step size with the accompanying "choppy" output, or use hardware that can provide greater ADC precision, thus allowing the same voltage accuracy but able to accommodate a greater input voltage range. Fig. 1 - 4 shows the influence which amplifier gain has on the dynamic range and voltage accuracy. The input consists of the sum of signal and noise. At a low gain, there is no saturation (or clipping), but the low voltage details are missing. At high gain (x5) these details are clear, but clipping has occurred. Such a compromise between gain and fidelity may be avoided if the original data had been sampled at a higher precision, effectively raising the clipping threshold. Like alias error, saturation error is permanent, with no recourse for data recovery once sampling is completed. If a fixed precision ADC is used, the only recourse is for the amplifier gain to be lowered until the dynamic range is sufficient to avoid clipping.

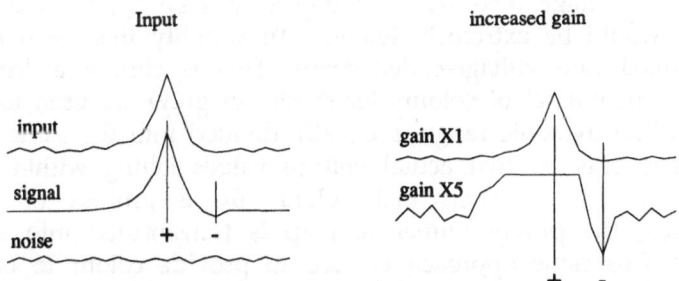

1-4:   Amplifier gain and clipping. Left: the input tracing (top) consists of signal (middle) and noise (bottom). Right: the effect of increasing the gain: the bottom tracing (x 5) shows clipping of the positive peak, while the smaller noise ripples are enhanced.

Map construction

Let us assume that a set of scalp electrodes applied according to the International 10 - 20 System (henceforth referred to as 10 - 20 system) had been used to record an EEG signal, and that adequate digitization had been carried out. The resulting data is in the form of individual tracings which represent voltage variation with time, or a "time series". One can display these tracings in anatomically meaningful arrangements, by arranging the tracings more or less in the respective head locations. To be able to fully appreciate at a glance all the subtleties of the spatial information present, a totally different form of data display is required. Such a display must convey the anatomic or spatial element (i.e., based on some sort of a head diagram), and also the amplitude information pertaining to the potential field of interest. Fig. 1 - 5 illustrates how the common example of frequency power display can be adapted, showing the topographic features of the various frequency bands. No ideal method exists that can allow all the temporal and spatial information of a segment of EEG data to be displayed in one diagram. By limiting each individual display to a given time window, and by allowing a successive series of such displays, useful topographic displays can be constructed, based on the idea that a pictorial representation of the instantaneous potential field can convey useful information.

A rudimentary "map" can be constructed merely as the arrangement of the voltage amplitude at one time instant, as given in Fig. 1 - 6. This is inconvenient for visual analysis, as all 19 voltage values in each "frame" must be completely scanned and remembered before the voltage field which it depicts can be appreciated. Such a process would be extremely tedious. To simplify this, each frame is transformed into voltage-coded maps. This is simply a "colouring" process where a set of colours (or shades of grey) are used to convey voltage. The dynamic range is equally divided into the same number of voltage bins, so that actual voltage values falling within a given colour's bin is assigned that colour. By a process of "pseudo colouring", the purely numerical map is transformed into a colour display. This same approach is used to provide colour to originally black and white motion pictures. Fig. 1 - 7 illustrates the process of map construction.

Fp1   Fpz   Fp2

F7   F3   Fz   F4   F8

T3   C3   Cz   C4   T4

T5   P3   Pz   P4   T6

O1   Oz   O2

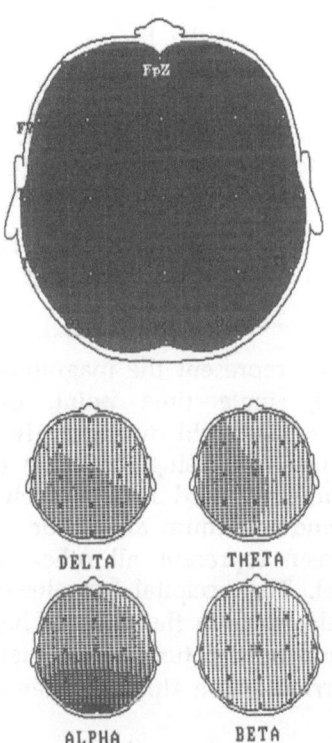

1-5: Map construction. The electrode array (top) is superimposed onto a head diagram (middle). By assigning appropriate values at each of the electrode positions, meaningful topographic displays result (spectral band maps - bottom).

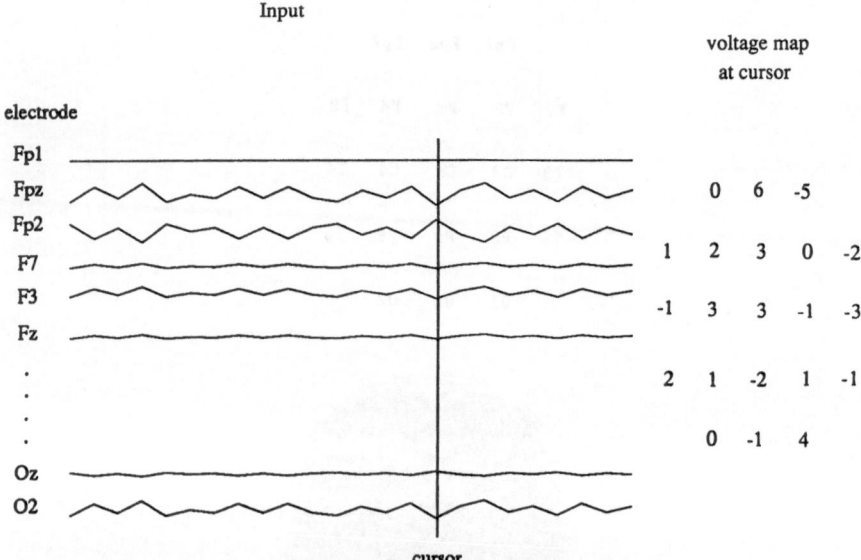

1-6:   Numerical display. The voltage readings at all channels at any given instant (cursor on the left) is displayed anatomically (right).

A single map may represent the magnitude of the instantaneous potential field for a single time point, or some mathematical representation of the voltage field during a given time window (e.g., mean voltage). An illustration using the alpha rhythm can be seen in Fig. 1 - 8. The colour scale used here is such that black and white represent maximum and minimum amplitudes respectively, while the grey shades in between represent all other intervening amplitudes (i.e., a unipolar scale). The occipital prominence is obvious, with a tapering-off of the field towards the anterior head. If sequential maps are shown, there would be a pattern of fluctuation between high and low magnitudes occurring at the alpha frequency of 10 Hz.

At the coarser time windows of 1 sec. used here, the details of how the field changes with time is lost, and all that remains is the overall pattern of occipital prominence. This happens since the original data is a persistent occipital dominant alpha rhythm, with no fluctuation or change over the time period under study. The maps for the 1 sec. windows thus depict the averaged activity over the

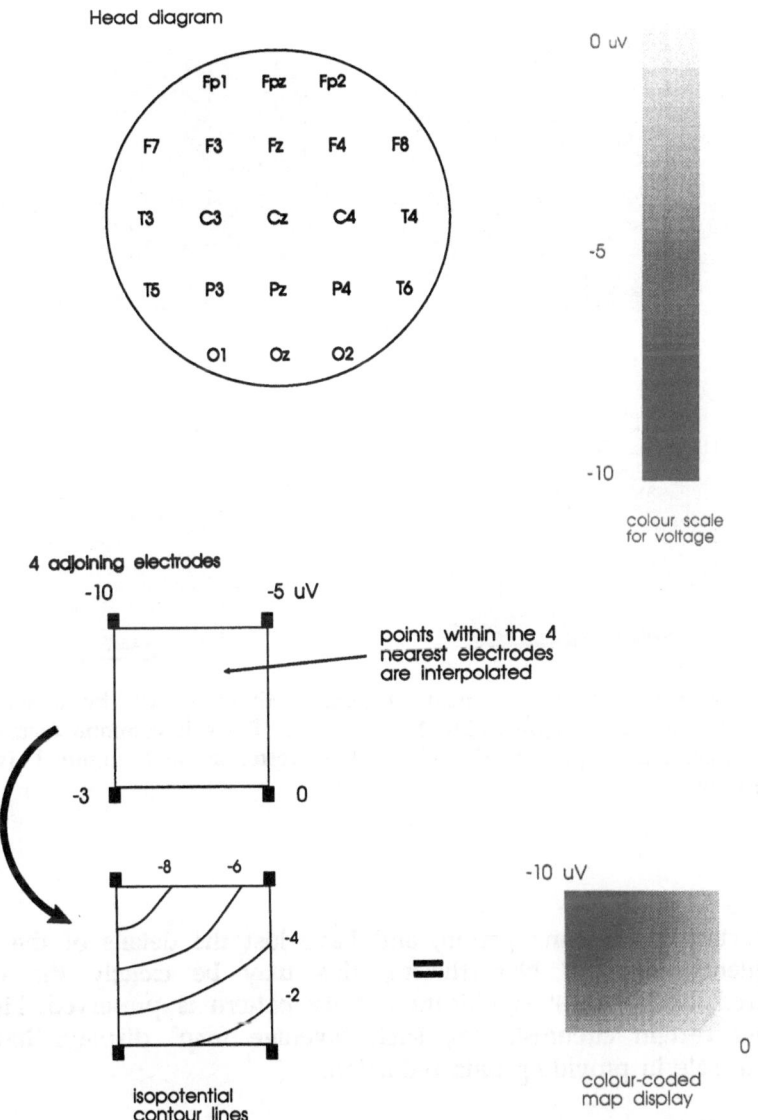

1-7: Map construction. Electrode array (top left); single square of 4 electrodes with an interior point (mid left); interpolated with voltage contour lines (bottom left); colour/grey voltage scale (top right); grey-scale representation of the interpolated square (bottom right).

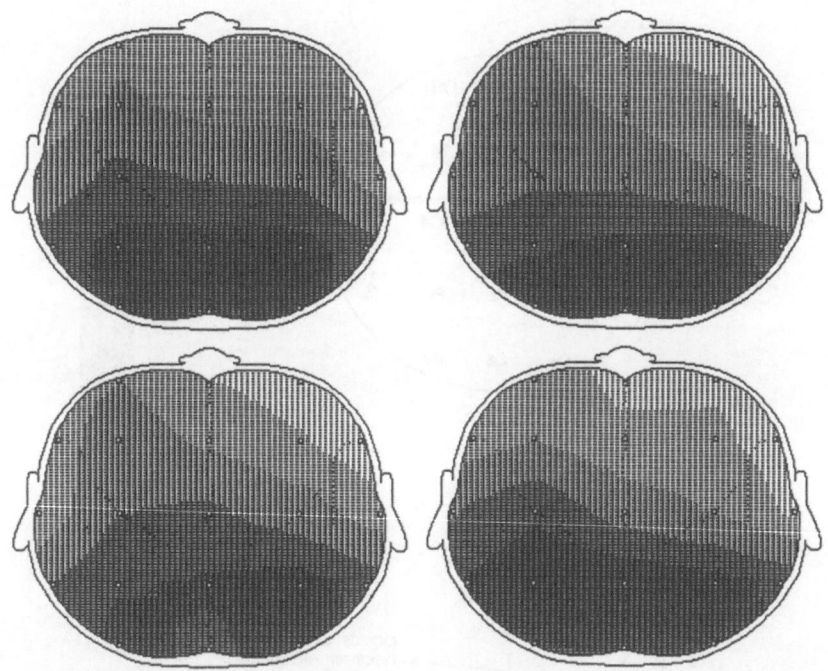

1-8: Map display. Simple field of alpha activity, shown by a series of average maps, each representing 1 sec of data. There is common field shape with maximal amplitude (black) at the posterior, and minimal (white) anteriorly.

respective 1 sec. time period, and have lost the details of the high frequency changes. Nevertheless, this may be exactly the effect desired, as the most significant activity pattern is preserved. Hence, under certain circumstances, such "average map" displays have a useful role in providing data reduction.

Interpolation

The method of interpolation used in map construction have a significant influence on the final appearance. A word of caution: it is not really necessary for the novice to pursue the mathematical details of interpolation at this point in order to assimilate the following

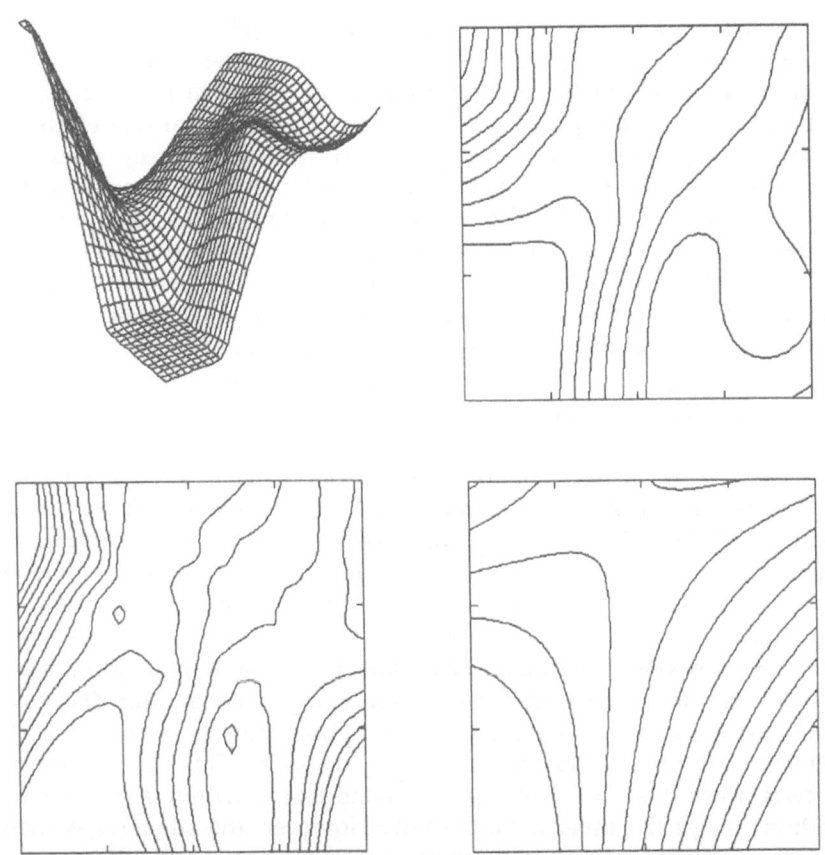

1-9:   Interpolation methods. Top left: desired (or reference) topography (vertical axis is voltage amplitude); the other 3 contour maps are derived from using different interpolation methods. The different degrees of spatial details is obvious: maximum at bottom left, minimum at bottom right.

chapters. This part can be deferred if desired. While there are many different mathematical schemes, in practice, the following methods are the most common linear - nearest 3 or 4 neighbour; non-linear curve fitting methods using quadratic or higher order equations; and inverse square weighting algorithms, which render a focal peak more "sharp". Examples are illustrated in Fig. 1 - 9.

The nearest 3 and 4 neighbour linear algorithms are easiest to understand as the interpolated values merely follow the trend set by the bounding 3 or 4 real electrode values. They both have the characteristic that any peak voltages seen in the resultant map must lie at the locations of the real electrodes, and never in an interpolated location. Thus spatial peaks will be seen to "jump" from one electrode to the next. In this sense the perceived spatial accuracy does not exceed that of the original data, being limited by the density of the real electrode array. In the examples of non-linear interpolation, spatial peaks can occur anywhere, as the interpolation depends on the curvature of the underlying potential field. Thus the perceived spatial accuracy can be higher. Whether this is in fact appropriate or even correct may not be obvious. Generally non-linear schemes are computationally intensive, which may result in more expensive hardware being required.

Fig. 1 - 10 shows a map sequence depicting a travelling spatial peak which might represent a conducted spike discharge. It should be appreciated that the map sequence covers a short time window (5 msec. between maps), in essence allowing us a microscopic view of the spatial details of the discharge. Such details could not be appreciated on paper and pen EEG systems, due to the limited frequency response, alignment error and other mechanical limitations. The inverse-distance-squared algorithm would display the discharge much more focally, based on the assumption that the recorded scalp voltages fall off as the square of the distance. This assumption is derived from the fact that volume conduction within a conductive medium obeys the inverse square law. However, the situation is more complex in reality, as the voltage attenuation with distance is governed not only by intracranial volume conduction, but also by conduction within the essentially flat layer of scalp tissue. The latter situation clearly does not follow the inverse square law.

With the exception of inverse squared weighting which is based on some debatable assumptions, there is little practical difference between the other interpolation methods. Further, it has been shown that under usual circumstances as might be encountered in clinical practice, most of the algorithms discussed above are equivalent (Koles

---

1-10:   Interpolation effect. Example illustrating travelling phenomenon across scalp, from C4 location (top left, going down), to Cz (bottom right), where some rotational or directional change has occurred. Black = negative, white = positive.

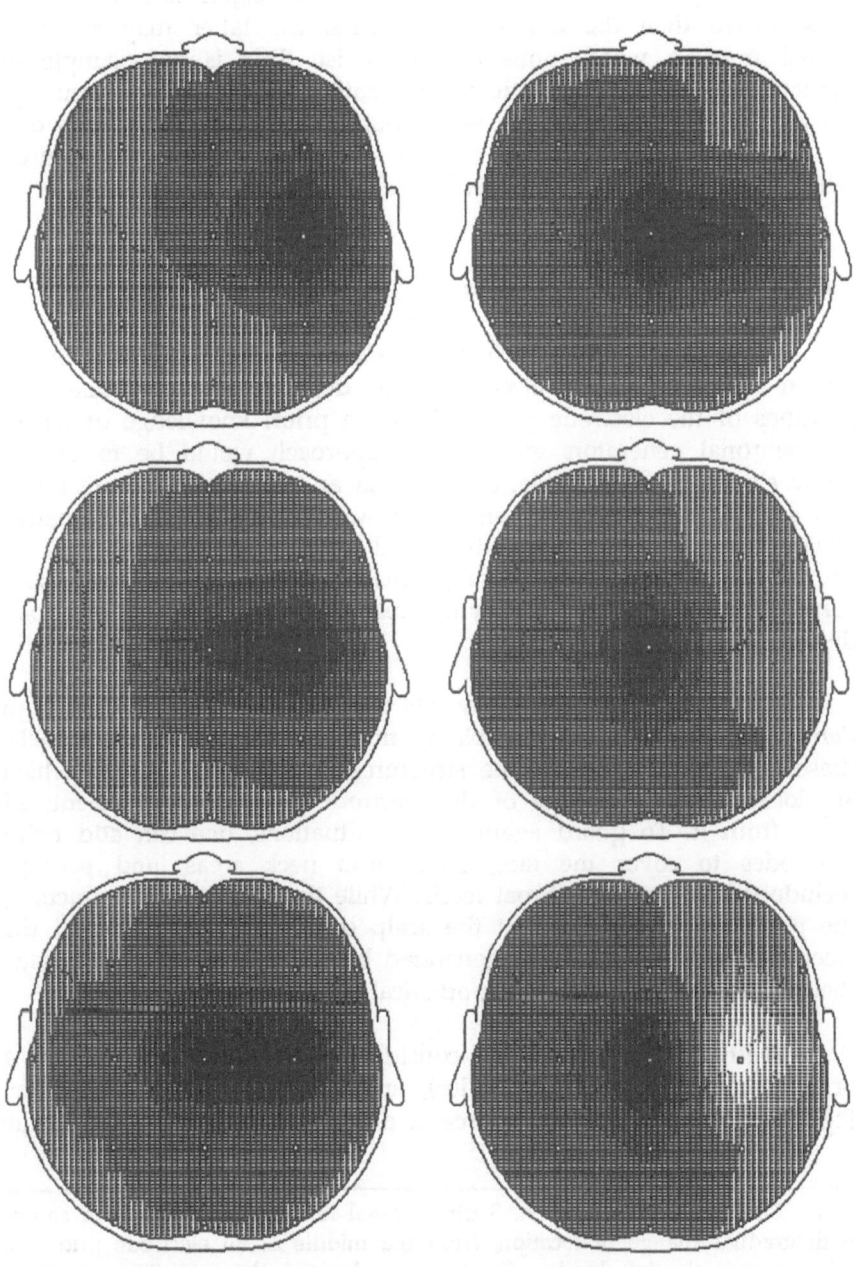

and Paranjape 1988). However, the linear algorithms are more conservative than the non-linear schemes: the latter may assign a spatial maxima where none actually exist. This is an example of perhaps unjustified optimistic representation of the spatial accuracy. As per the discussion in the next chapter, there are limitations due to limited spatial sampling density, therefore caution is required when attempts are made to predict scalp values.

1.3 SPATIAL SAMPLING

The accuracy with which one can localize the origin of a signal within a particular scalp electrode array depends on the number and locations of the electrode sites. Without a priori knowledge of where the neuronal generators are, a good approach would be to use as many electrodes as possible, arranged so as to cover the entire scalp, and spaced as evenly as possible. This would yield an approximately square array. The interpolation and display algorithms must of course take into account these geometric details of the array, and correct for any peculiar idiosyncrasies (e.g., different inter-electrode distances).

Even if a given scalp array has sufficient electrodes (i.e., a high density), it is still quite possible to miss certain intracranial signals: those deep within the midline structures, or neuronal sources which are located near the edge of the electrode array, and/or orientated away from it. To guard against these situations, one can add more electrodes to cover the face, cheek, and neck areas, and perhaps include sphenoidal and throat leads. While the latter are not practical, the point to be made is that the scalp covers only a portion of the electrical field which can be generated by intracranial neuronal tissue (henceforth referred to as a hypothetical source or simply source).

To illustrate these points, consider an array consisting of 5 x 4 or 20 recording points (electrodes), arranged in the 10 - 20 system. Fig. 1 - 11 shows a signal source at different locations relative to the

1-11: Effect of source location. 3 dimensional line figure display of a source with gradual change in location, from the middle of an electrode grid (top left) towards the left border, finally going beyond the grid (bottom right). Height = amplitude.

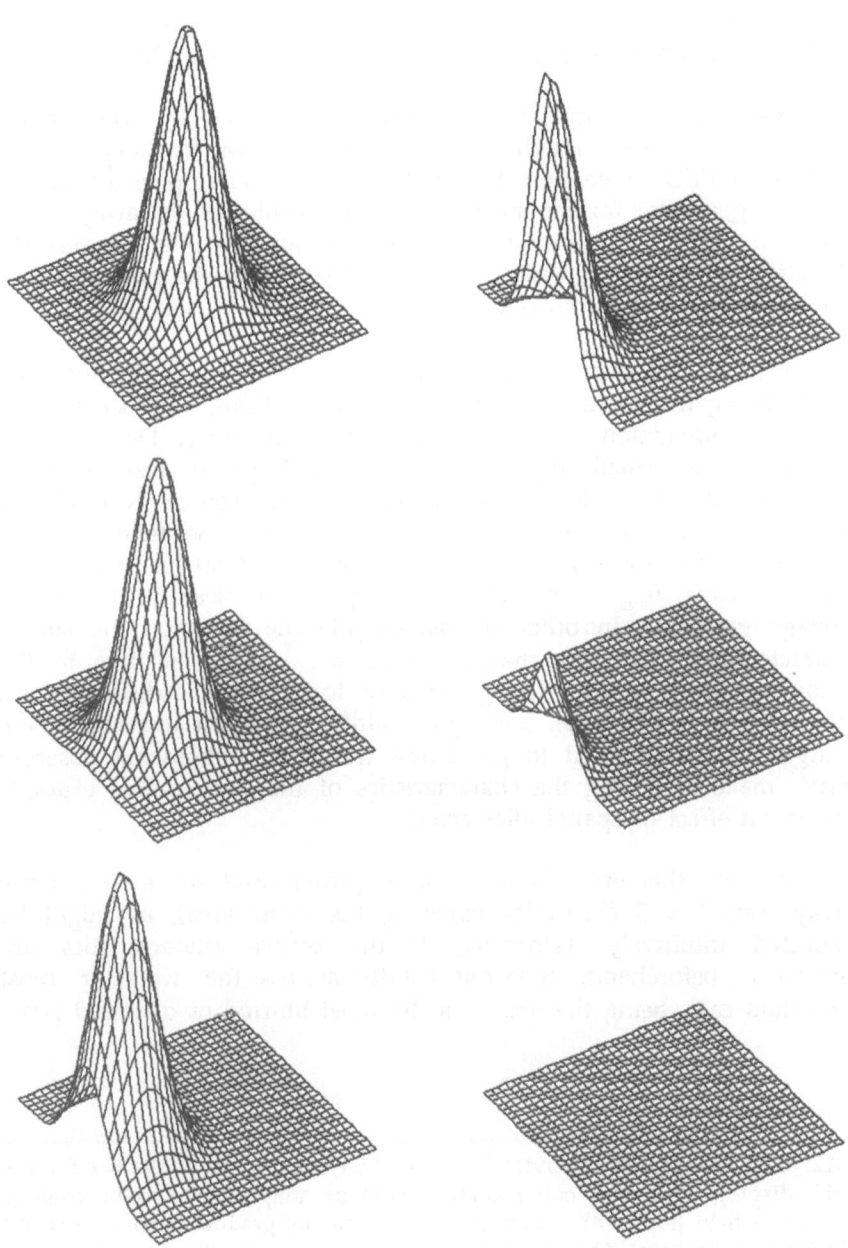

array, together with the respective readings at the peak (or maxima) of the signal. All amplifiers are recording in the referential mode, with the reference being assumed to be an appropriate choice.

The recorded topography clearly is influenced by whether the array is extensive enough to cover the field of the source. By traditional EEG convention, the field in Fig. 1 - 11 (top left) indicates that the generator for the signal is near the center of the array, while those in Fig. 1 - 11 (bottom or middle right) can only suggest the direction where the generator is, as the peak is off the grid and thus knowledge of it is incomplete.

Fig. 1 - 12 demonstrates more subtle changes in the recorded field arising from small changes in source locations, even though the source is adequately covered by the electrode array. Distances are referred to as small (e.g., 1 or 2 cm) if they are less than the interelectrode separation (5 or 6 cm). As the source is gradually relocated from directly under the C3 electrode, the field changes. Notice that there are certain locations where the field has a relatively focal maxima (e.g., sharpest voltage peak or peak with highest voltage gradient). In other words, despite the fact that the source characteristics have not changed, only the location relative to the recording array has, its voltage field or topography is different, and is not easily predictable. Such uncertainty causes an increase in the margin of error related to prediction which an unknowing observer may make regarding the characteristics of the source. This example shows an effect of spatial alias error.

In fact, this error is even more pronounced for a less dense array (say 3 x 3 electrodes covering the same area), as might be expected intuitively. Generally, if the source characteristics are unknown beforehand, then one has to assume the worst or most uncertain case, being the one with the most blurred or distorted peak.

1-12: Field changes with source locations. Left column shows the 6 x 6 array grid display, with the corresponding contour map in the right column. Reference field (top) with source at the C3 location; gradual movement to the C4 location (bottom). Orientations of all maps are generally anterior up, and right side being right, unless otherwise stated.

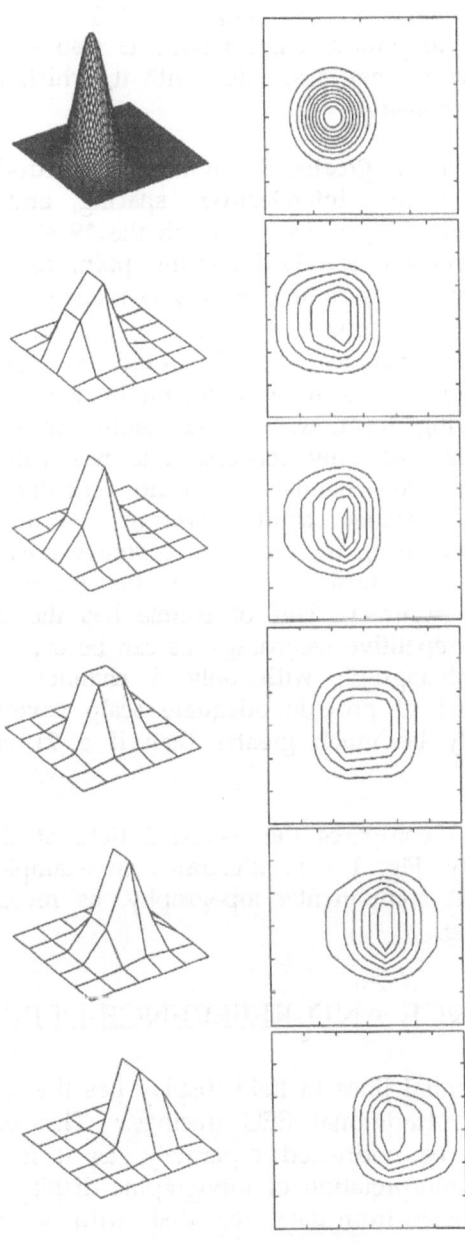

If one is interested in the location of the source alone, this corresponds to an error of approximately +/- 1 interelectrode distance. If the amplitude characteristic is also of interest, that too will have an error range associated with it, which is directly related to the electrode density.

Clearly then, the greater the number of electrodes used to cover the scalp, the smaller interelectrode spacing, and the greater the spatial resolution. In practice, although the 19 electrodes of the 10 - 20 system provides a standard starting point for clinical screening purposes, this spatial sampling density is inadequate for the accurate localization of many focal activities. If one is interested in the occipital area for instance, then it would be more appropriate to concentrate all electrodes in the posterior head region. Anything less than 19 recording locations for the entire scalp will give poor performance, and the only recourse one has with such a limited number of recording channels is to do repetitive recording runs, starting with a sparse array covering the entire scalp, then progressively decreasing the electrode coverage area into smaller and smaller scalp regions near the focus, to be able to define the signal with increasing accuracy. This of course has the familiar problems associated with repetitive recordings, as can be experienced if one has an electroencephalograph with only 8 channels. The number of montages needed to provide adequate scalp coverage in this case would obviously be much greater than if a 21 channel unit was available.

Fig. 1 - 13 compares the recorded field at different levels of electrode density. Fig. 1 - 14 illustrates an example of a simulated evoked potential component's topography, as revealed by different electrode densities.

1.4 REFERENCE AND REFERENCE-DEPENDENCE

The isopotential lines in field display has the same relation to a reference as the traditional EEG tracings. This very fundamental point needs to be addressed separately, as it is a key principle underlying the interpretation of topographic displays. Recalling that a map is constructed from data recorded with a common electrode

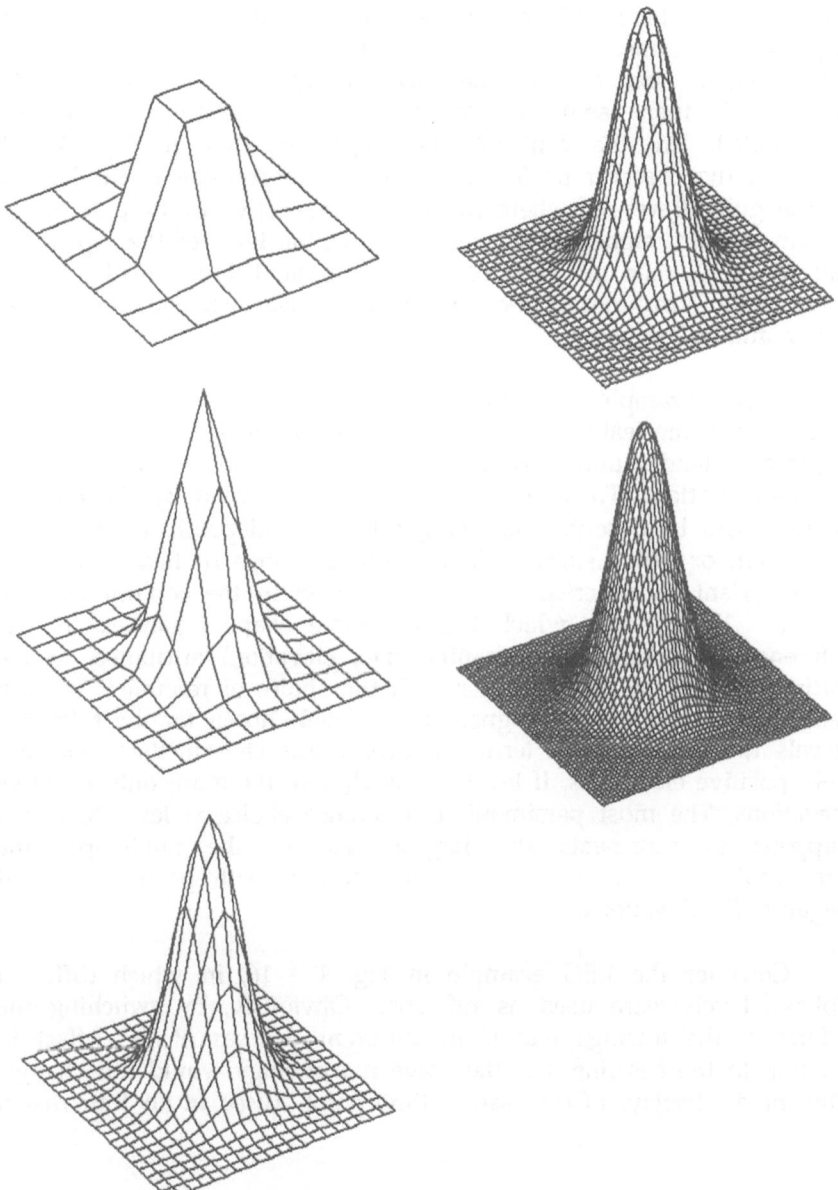

1-13: Electrode density. Arrays of increasing density, from 6x6 (top left), 11 x 11 (mid left), to the most dense 128 x 128 (bottom right).

as reference. There will be voltage swings in the more positive and negative directions. If one takes the analogy of a contour map of a hill situated next to an equal-sized gully in the ground, the reference is then analogous to the location where the observer is situated. In the artificial landscape depicted in Fig. 1 - 15 (top), let the observer be first at a level corresponding to the bottom of the gully (level 1), where from this perspective every part of the terrain appears higher. If he were to be at the level of the top of the hill (level 2), then all parts of the terrain lie below. Finally, at the level of the surrounding plain (level 3), the view shows some hilltop above and valley below.

In this example, it should be noted that regardless of the point of reference, the features of the terrain do not change. Hilltop is still a piece of land jutting upwards, while water tends to run down to the gully bottom. These invariant features (governed by the laws of physics) can be described as being reference-independent. The shape of the hill or valley surface, their angularity slope or flatness etc., all are invariant characteristics. This is reflected in the contour map in Fig. 1 - 15 (bottom), which has the same shape regardless of the reference level. Only the meaning (i.e., elevation) attributed to the individual contour lines changes with the choice of reference. In each case, "zero" elevation is assigned to the level chosen for the reference (levels 1, 2 or 3). In real terms, if level 1 was chosen, then there are only positive elevations. If level 2 was chosen, there are only negative elevations. The most parsimonious reference choice is level 3, which happens to represents the largest area of the landscape (the surrounding plain), and which will result in both positive (2) and negative (1) elevations.

Consider the EEG example in Fig. 1 - 16, in which different voltage levels were used as reference. Obviously, by switching the reference, the tracings will show different degrees of DC offset in relation to the baseline, but the wave morphology would not change. The field displays of the same time point (used in all four maps)

1-14: Spatial sampling.   A simulated EP component's topography using different array densities similar to Fig. 1-13. Positive up. There is a large positivity at the left anterior quadrant and large negativity at the left posterior.

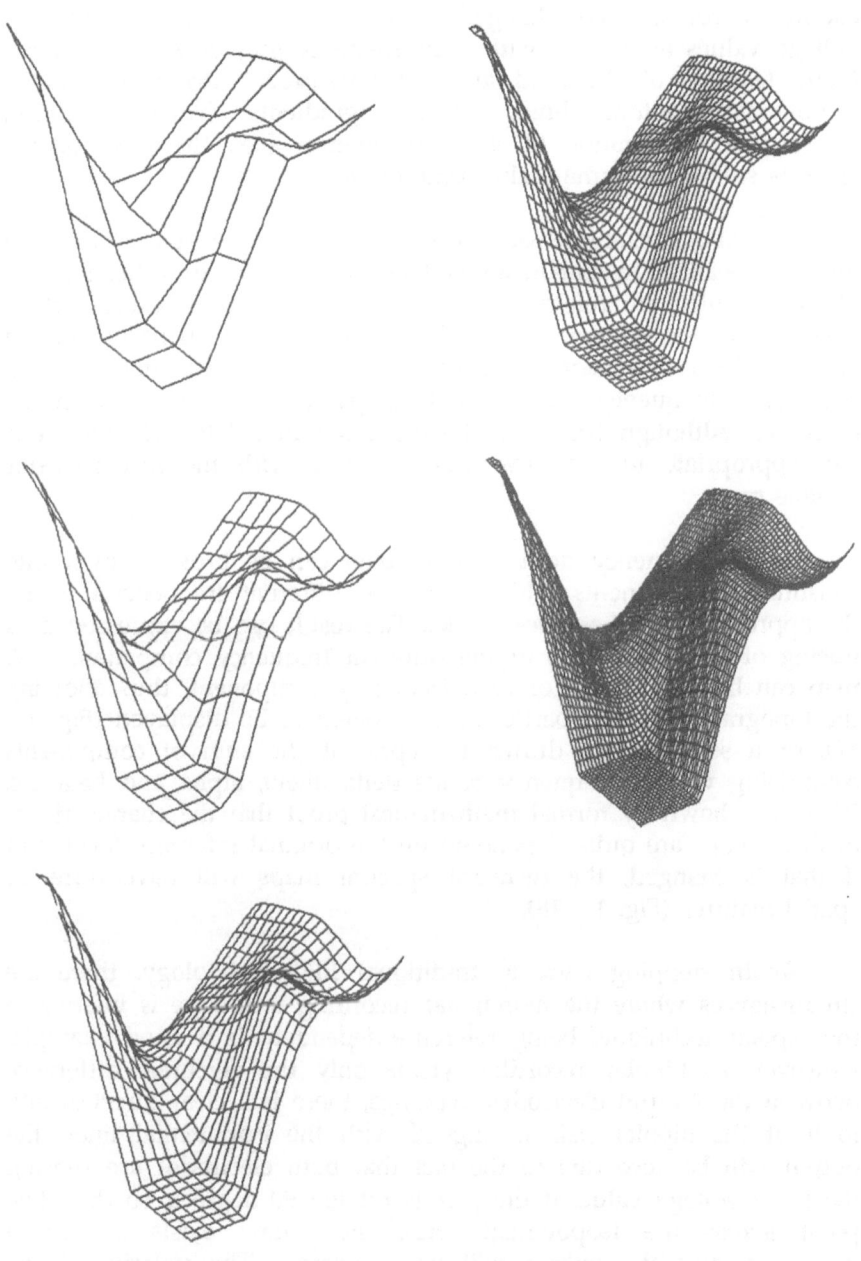

show identical contour lines, although the grey shading (or colour coding of voltage) has changed in meaning. The result is that the voltage values associated with each shade of grey had been altered. Many features of the field are still preserved: the shape of the isopotential contour lines, voltage gradients (voltage slopes), topographic distribution of items of interest (i.e., the peak of the spike is still at the same scalp location) etc.

The above example contained voltage maps taken at a given instant, so-called "instantaneous voltage map". There are other families of maps which are reference-dependent, those whose features which change with the reference point. These are generally maps of derived values, which had been transformed using non-linear mathematical operators. Frequency or spectral maps are the most common examples. Although these will be discussed in a later chapter, it is still appropriate to see how they compare with the instantaneous voltage maps.

In the frequency domain, EEG data can be broken down into constituent components which represent different frequencies, using the appropriate mathematical tricks. The result can be expressed as a tracing of the amplitudes of the different frequency components. A map can be constructed of each frequency component, thus allowing the topography of that particular component to be displayed (Fig. 1 - 17); or a set of maps drawn to represent the sum of components comprising various frequency bands delta, theta, alpha and beta etc. It can be shown by formal mathematical proof that the characteristics of these maps are quite dependant on the original reference level, and if that is changed, the resultant spectral maps will have different spatial features (Fig. 1 - 18).

Again stepping back to traditional EEG technology, there are circumstances where the monopolar recording technique is inferior to the bipolar technique, being reference-dependent is a good example. However, as bipolar recording yields only the algebraic difference between the 2 input electrodes' readings, there is a clear directionality to it. If the bipolar pair is aligned with the isopotential lines, the output will be zero due to the fact that both electrodes are sensing the same voltage value. If the pair is rotated 90 degrees, so that they point across the isopotential lines, there now exists a voltage difference, and the output will be non-zero. The polarity of the output is dependent on the direction of the pair: it will be reversed

REFERENCE LEVELS

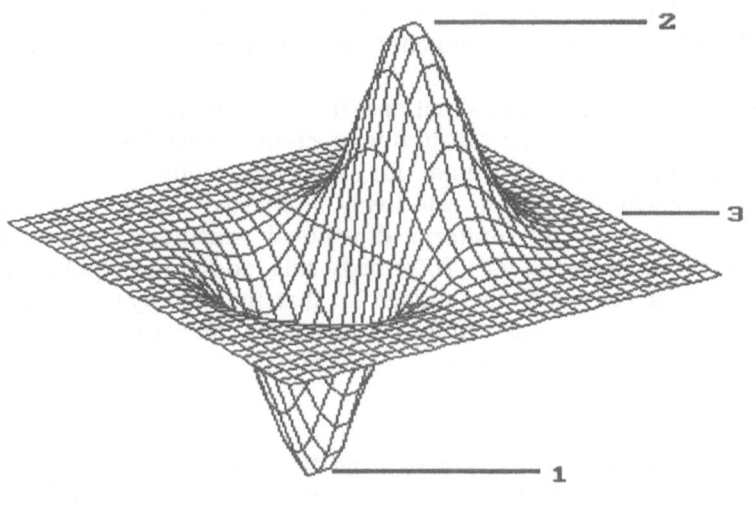

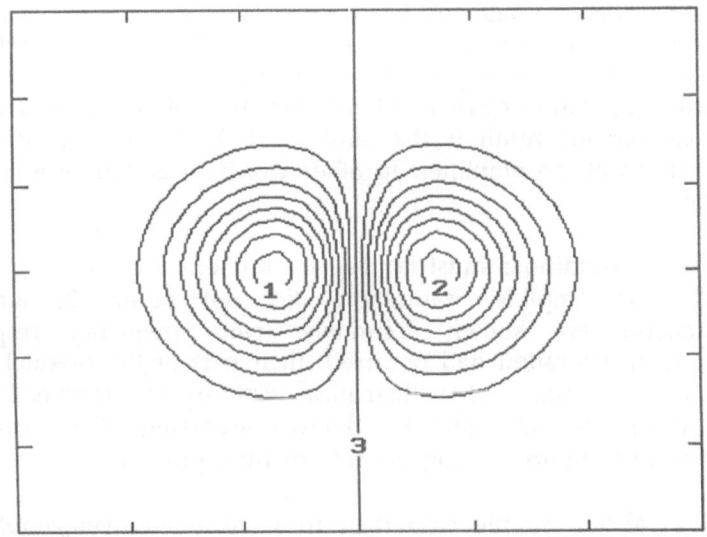

1-15: Topography and reference: 3 levels to look at a terrain with a hill and gully. Top: contour map with the 3 reference levels indicated, where level 3 is that of the surrounding ground (plain). Bottom: the corresponding contour map, with the left side being lower and the right being higher than the plain (level 3). Here each contour line has been a assigned specific height value.

if the pair is rotated 180°. From this simple example, it can be demonstrated that the output of this bipolar pair is a function of the angle it subtends with the isopotential lines ("directional sensitivity").

In real-life complicated situations, the potential field lines are not simple, and the directional sensitivity function is complex. Whereas a bipolar montage can be used for mapping, with its attendant advantages including being selectively affected by local voltage changes,   it lacks the uniform directional sensitivity that the monopolar technique offers. To remedy this, additional bipolar pairs are required at the same site, aligned so that the aggregate sensitivity function of the resultant triplet is more or less uniform in all directions. Fig. 1 - 19 illustrates the directional sensitivity functions of these different recording methods.

Bipolar electrode triplets have been used in recording probes to detect myocardial signals *in vivo*. It does triple the number of channels required for each recording point. For the 10 - 20 system's 19 positions, this means 57 recording channels. For this reason, bipolar montage is not often used directly, but created or reformatted from the original data obtained with the monopolar technique. Shewmon and Krentler et al. (1984) described this process. Briefly, the digitized data is manipulated in pairs, and the mathematical subtraction operation carried out for each pair of the desired bipolar chain, so that the result is the same as if the subtraction of signals had occurred at the amplifier. In other words, it is merely a software operation.

Certain conditions must be true for this operation to give correct results: 1) amplifier saturation did not occur; 2) amplifier characteristics are known accurately (gain, frequency response). Obviously if saturation had occurred, then part of the original signal is lost, and no amount of mathematical wizardry will recover it. Even if the gains are not equal but known accurately, then correction factors can be applied during the reformatting process.

It is also a simple step then to recreate an average-reference dataset from the original referential dataset. The computation of the mathematical average of all sampled electrode readings is used as the reference and subtracted from all individual electrode's readings. Such a procedure need to be done at each time point, not any hardship for

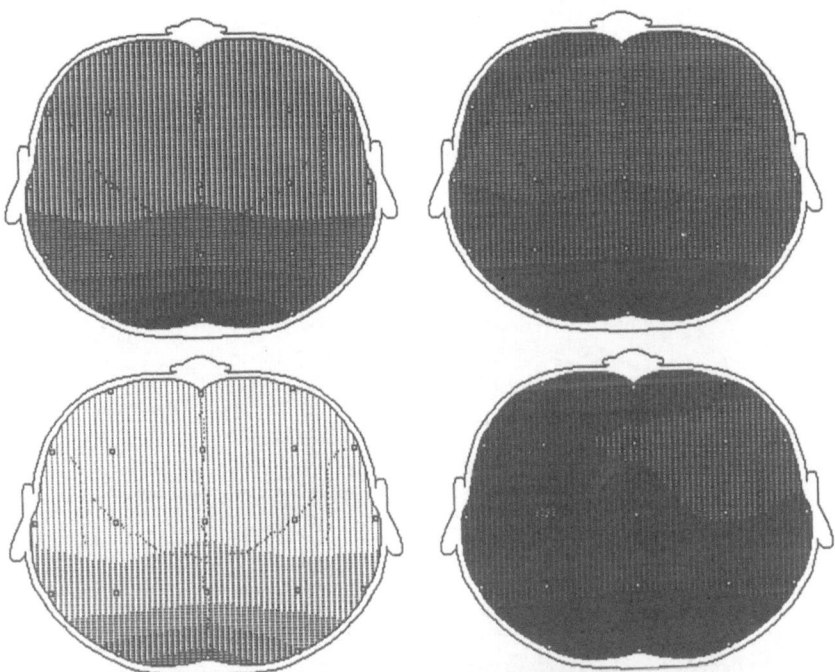

1-16: Effect of reference: 4 maps drawn with different reference levels. Minor variations are due to mathematical rounding errors (this is noticeable only in the bottom right map). The main field outlines remain constant: there is always an occipital prominent feature, and relatively flat gradient in the anterior head.

today's microcomputers. Fig. 1 - 20 shows the same dataset using monopolar and average-reference technique.

This is an example of post-hoc softward processing, which yields useful data transformation without special hardware. Such "post-processing" capability can be very simple (like reformatting), or extremely complex. It is discussed in greater detail later.

1.5 MAP DISPLAY METHODS

It is instructive to explore some of the ways in which the same data can be topographically viewed. Given a frame of data consisting

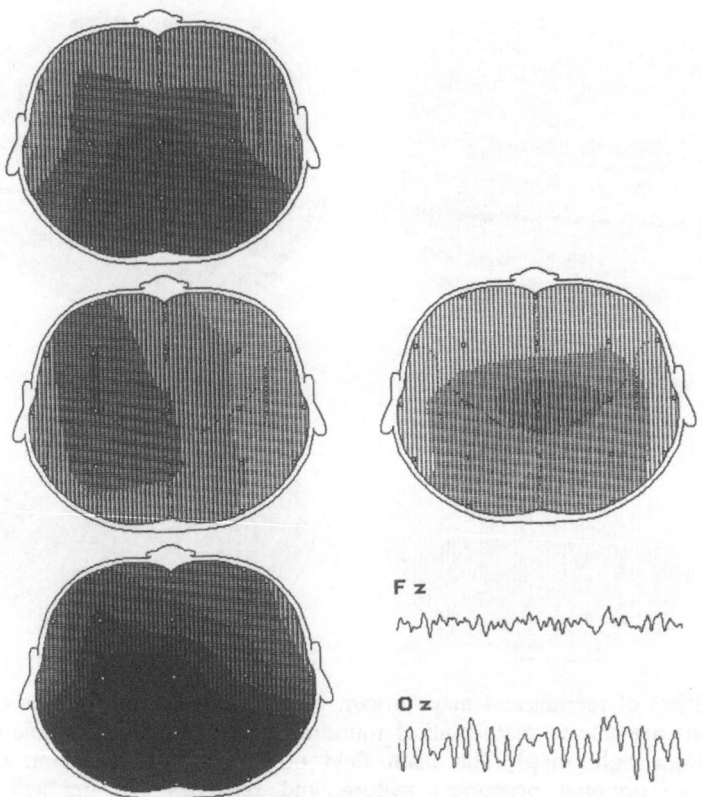

1-17:   Spectral maps at different frequencies: 2.5 Hz (top left), 5.5 Hz (middle left), 10 Hz (bottom left), 18 Hz (middle right). Raw EEG tracings at the Fz and Oz electrodes (bottom right).

of a set of 21 numbers, we can look at it with different display methods and from various visual perspectives. The following display schemes will be illustrated: contour line, contour line with shading, grid lattice and gradient vector.

Fig. 1 - 21 shows a frame of EEG data measured with a set of

1-18:   Reference effect: 3 different reference levels used in the raw data which were used to derive the 3 spectral maps in the same way as Fig. 1 - 17. Although the occipital prominence is present in all 3 (due to the overwhelming effect of the alpha activity being present), the other field lines are different. If more shades of grey were available for display these field differences would be enhanced. Compare with Fig. 1 - 16.

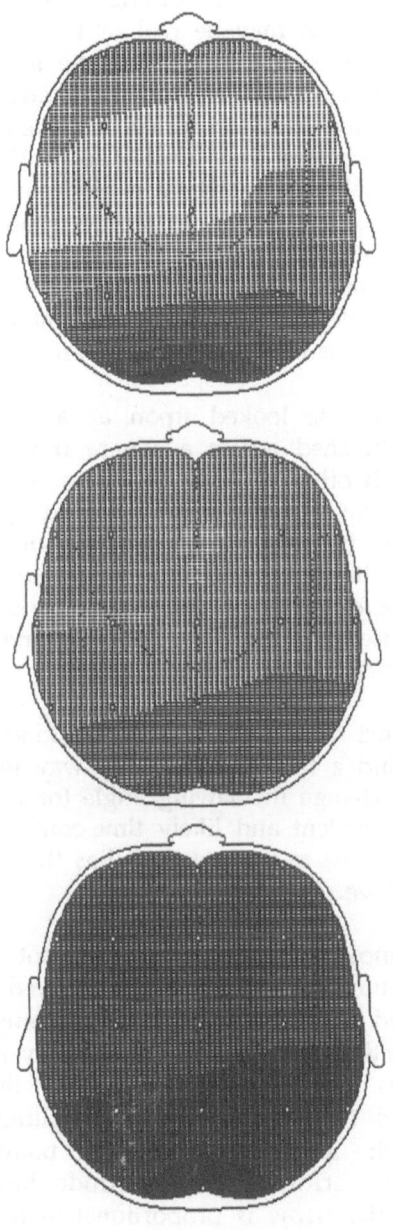

21 electrodes. The potential field can be inspected using the simple contour line map (middle). This is nothing more than a line drawing connecting locations (which may be real or interpolated from the 21 real values) which shows the same voltage reading. These lines arethe isopotential lines, and each is associated with a definite voltage value. In practice, it is sometimes inconvenient to display the actual number with each line, thus leading to possible confusion as to whether the isopotential lines represent a hill (positivity) or an abyss (negativity). A simple remedy is to apply shading to either the positivity or the negativity. A more refined remedy is to assign a different shading intensity to each isopotential line, and displaying the voltage-shading scale nearby. This is of course the now-familiar topographic display we had been using (bottom).

Grid displays can be looked upon as a set of rubber bands (isopotential lines) stretched across a square frame (representing the head), parallel to each other in both directions. A positive voltage at a particular location would pull the rubber band up, while a negative voltage would push it down. The amount of travel is directly proportional to the voltage magnitude. In this model, there is no need to interpolate, as the elasticity of the rubber band smooths out any abrupt changes between electrodes. Harris and Bickford (1968) described a similar mapping display method.

There is a distinct disadvantage with this kind of map: features may be hidden behind a hill, and the only way to visualize it is to rotate the map by a change in viewing angle (or perspective: see Fig. 1 - 22). This is inconvenient and likely time-consuming, as the entire graphical image has to be re-computed. It has the advantage of being naturalistic and intuitive.

There are instances when the display is not of the actual raw data but other computed or mathematically derived variable. An item of interest is the field gradient, which is more complex in that it has amplitude and direction information at any given location (i.e., a vector). It is obvious that to display an entity this complex, more ingenuity is required than displaying an amplitude. The following scheme may be used: an arrow is used for pointing the direction, with the base of the arrow at the electrode location. The length and/or thickness of the arrow is proportional to the voltage gradient magnitude, which can be calculated as the net voltage difference

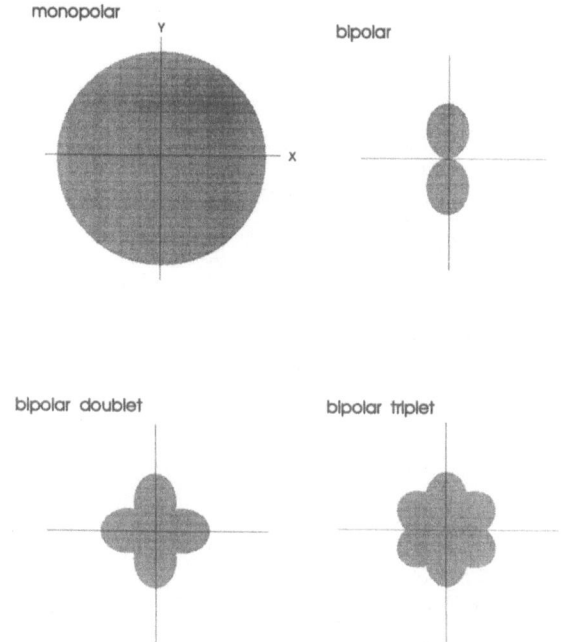

1-19: Directional sensitivity of monopolar and dipolar electrode pairs as a function of the angle of the recording electrode pairs. The monopolar case has a large even area of sensitivity (omnidirectional). The bipolar derivation is sensitive to local potential gradients; a single bipolar pair (top right) is clearly bi-directional, while a combination of 2 (doublet) or 3 (triplet) such pairs is required to approximate omni-directional sensitivity. Doublets and triplets are arranged with a staggered orientation for each pair (90° and 60° resp.).

between adjacent electrodes in both the anterior-posterior and right-left directions (Fig. 1 - 23). The resultant vector map depicts which way and how strongly electrical current flows on this surface. Further mathematical manipulation yields diagrammatic representation of the scalp patterns of current flow including the locations of outflow and inflow of electrical current ("source" and "sink"), thereby suggesting locations of the underlying generators. Such current source density maps provide additional insight in certain complex situations, and can be approximated by simple mathematical transformation of the raw data.

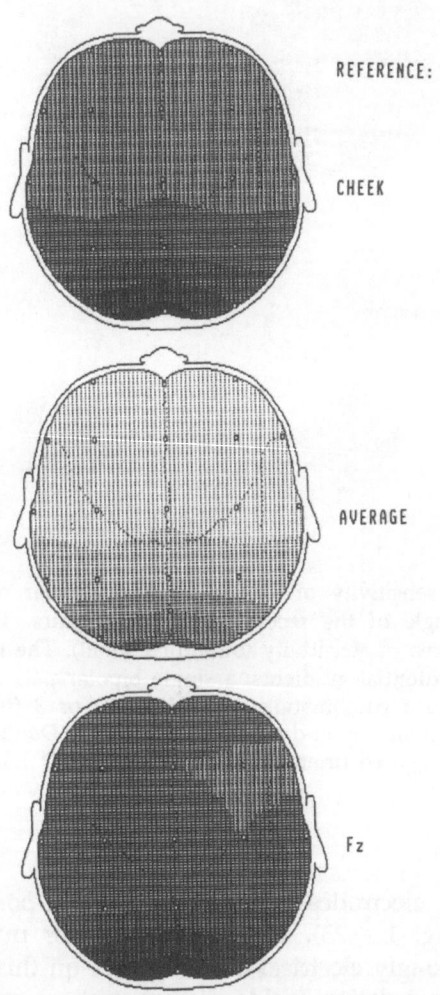

REFERENCE:

CHEEK

AVERAGE

Fz

1-20: Monopolar and average reference. Different references used to display the same EEG data. Note the similar field shapes.

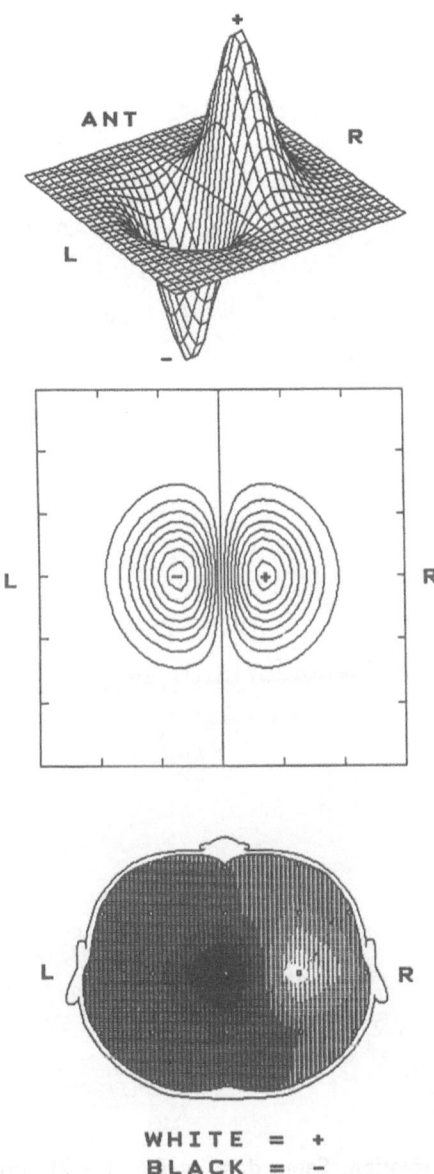

1-21:   Different ways to map data. Top: grid display; middle: contour map (vertex view); bottom: topographic map of head (same vertex view).

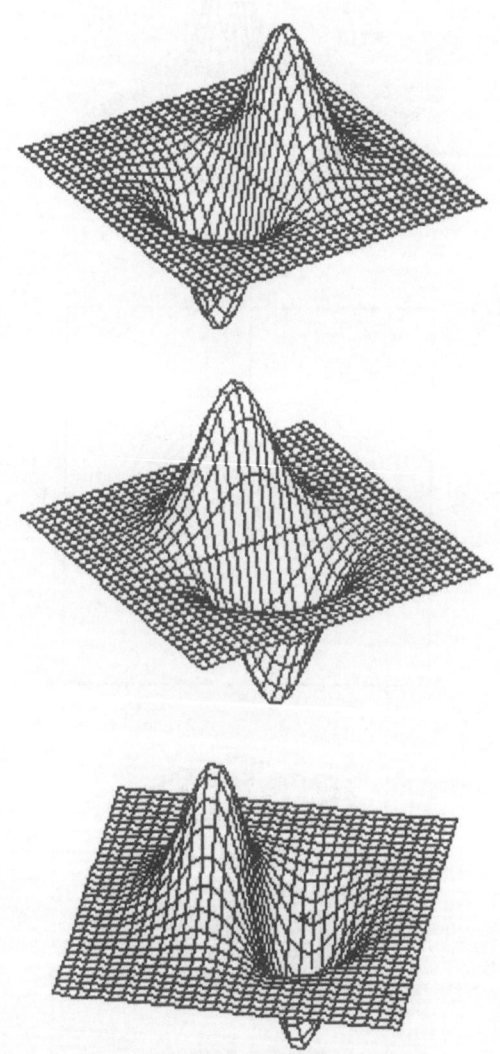

1-22: Grid display rotation. Same data as Fig. 1 - 21 top, rotated to show hidden surfaces.

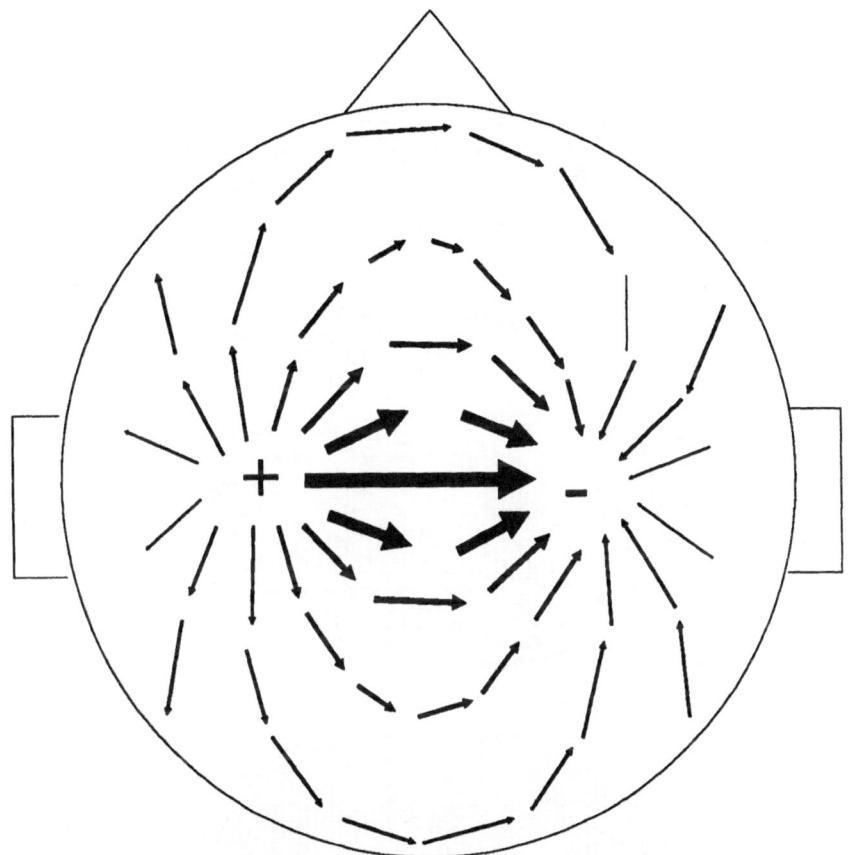

1-23: Vector map of scalp current flow. Same data as Fig. 1 - 22, showing the current flow from the positive "source" (C3) to the negative "sink" (C4).

Hjorth (1975) described the "source derivation" procedure, which is in essence a reference calculation such that each electrode is referred to the weighted aggregate sum of its neighbours. The net effect is that each electrode is sensitive only to local activity, distant activity being attenuated. There is thus a spatial high-pass filter action. It had been shown in clinical EEG that diffuse delta activity is not accurately portrayed, while focal activity (spikes etc.) are rendered more clearly.   Fig. 1 - 24 shows the topography at the peak negative

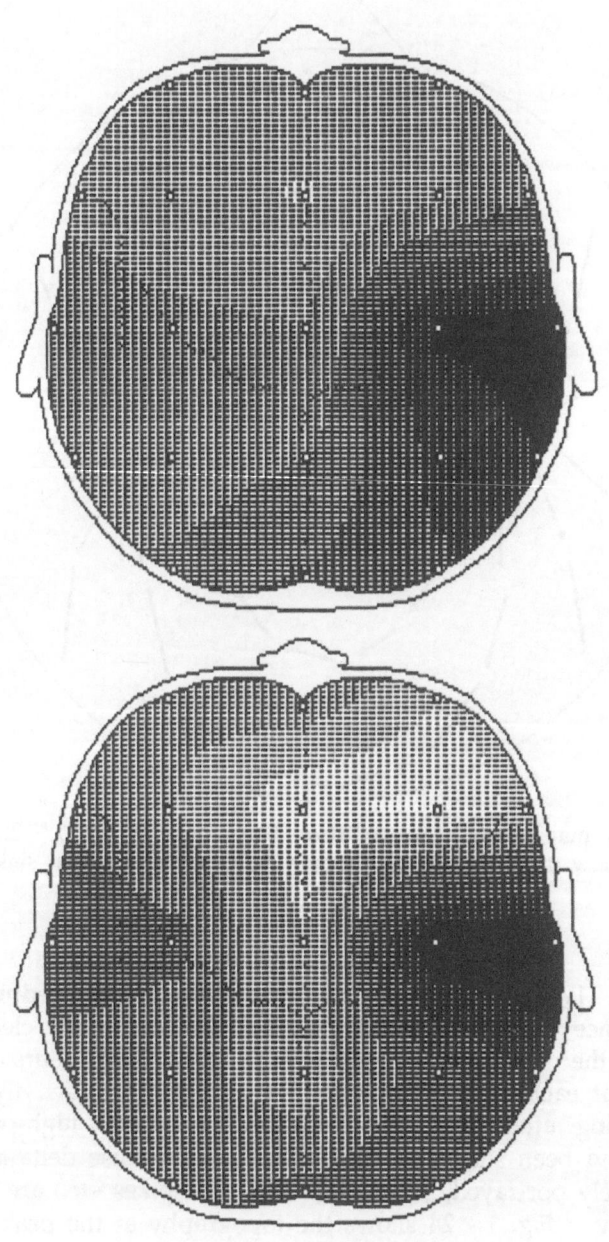

apex of a right central-temporal spike displayed in both monopolar (cheek) and source derivations. The former has a widespread positivity across the bifrontal regions, while the latter more clearly reveals the temporal-central origin of the spike.

Scaling and floating voltage scales

The assignment of colours to voltage values described earlier is the more commonly encountered than shades of grey. A given voltage range is unambiguously described. One may change the display scale, usually for convenience by integral powers of 2:   x2, x4, x8 etc. This then requires the 2:   x2, ... display scale to be expanded (Fig. 1 - 25). Such display scale changes is usually done to increase the size of low amplitude features, even though it may cause saturation of the higher amplitude features.

One way to allow maximal utilization of the colour range available is to use a floating colour scale; such a scale always assigns the first colour (e.g., black) to the least positive (or most negative) voltage, and the last colour (e.g., white) for the most positive voltage of the set of 21 values in a given frame to be displayed. The result is that the colour scale is customised for each frame, so that a given colour (e.g., medium grey) has a different voltage meaning for each frame.   While it does make best use of the limited colours, there is the unavoidable ambiguity of what a given colour actually represent in a location of interest. In a fixed-scale map, a given level of grey (e.g., medium grey) always represent the same voltage range.

Summary maps

Summary maps are available which provide varying degrees of data reduction. For example, if it is desired to inspect the locations where voltage maximums occur within a given time window ("epoch"), a "maximum" map can be drawn from the existing data. This is usually not very interesting, as a voltage maximum can be due to an artifact (muscle twitch) or a large peak.   However, a better

1-24:   Source derivation. Top: referential map showing right temporal negativity (black) with simultaneous left-mid frontal positivity (white). Bottom: Hjorth or source derivation map; note the more concentrated and enhanced activity around the focus at the right frontal-temporal area.

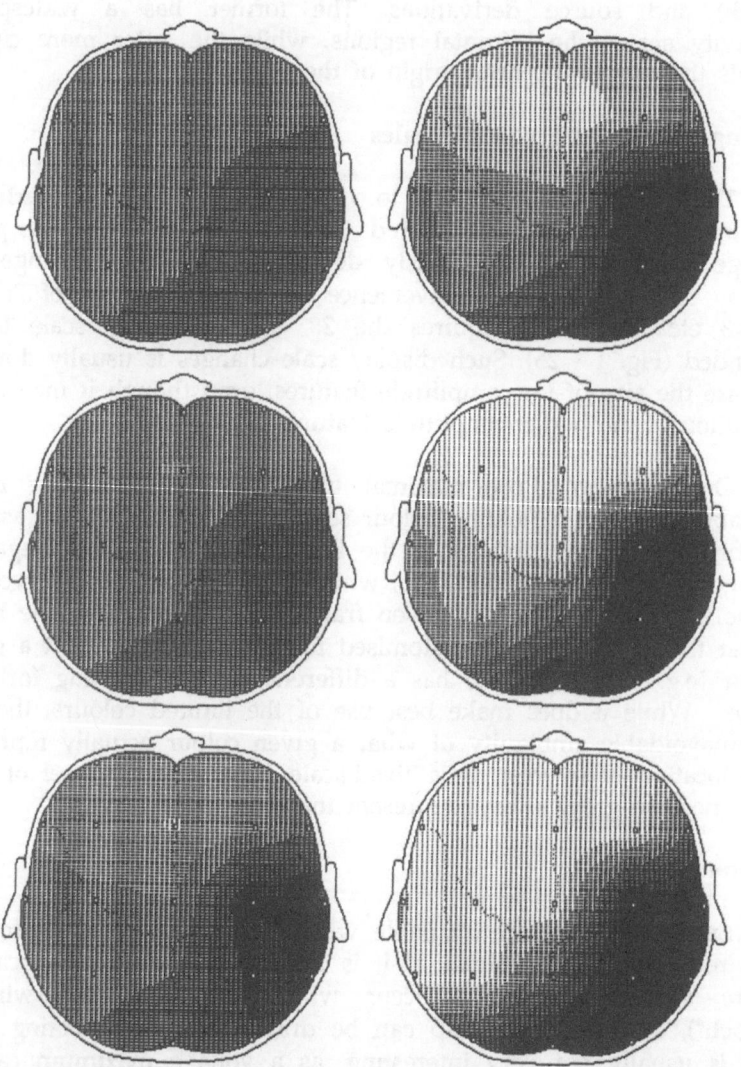

1-25: Display scaling. Maps displaying the same data as Fig. 1 - 24 but at
different scales: top left: x1, thereafter increasing by a factor of 2 from top left
down to bottom right (x32). There is increasing detail at x2 and x4 (bottom
left), saturation at the right posterior quadrant and progressive loss of details
of the left frontal areas from x8 (top right) onwards.

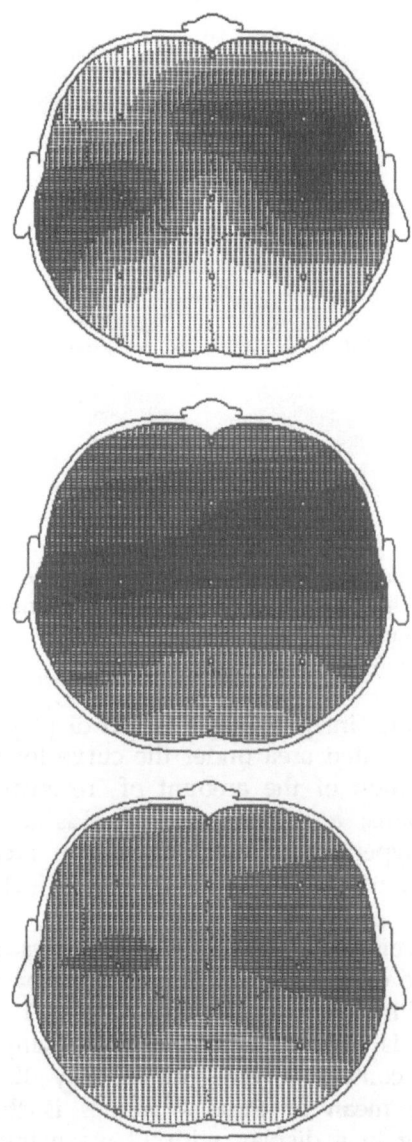

1-26: Summary maps: VEP data summarized in different ways over a 0.5 sec window. Top: maximum voltages at a given electrode (maximum map), showing greatest voltages posteriorly (white = +, black = -). Middle: area under the curve (unsigned integral map), showing posterior dominant activity regardless of polarity (white = maximum, black = minimum activity). Bottom: signed integral map, showing the posterior positivity within this window.

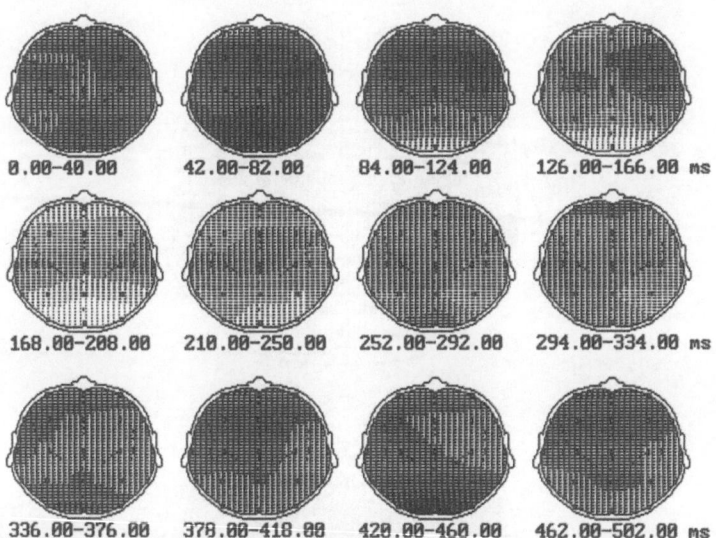

1-27a: Summary maps: 12 summary VEP maps (signed integral) covering the epoch 0-500 msec. White = +, black = -. There is prominent positivity at the occipital area at around 200 msec.

way is to inspect the "integral map", which displays the summated or mathematically integrated area under the curve for the epoch. This is a crude approximation of the amount of "reactivity" of the brain at various scalp regions; a greater area   yields a whiter colour and signifies relative hyperactivity, while a smaller area yields a blacker colour and signifies relative hypoactivity (Fig. 1 - 26).

Another convenient approach which allows rapid scanning of data is the average or mean map. This is done by selecting an epoch which might be of interest (say 0-500 msec., Fig. 1 - 27a). By software action, this epoch is further segregated into many smaller windows which may each contain several frames (say 12 windows with 4 frames each). The mean of these 4 frames is then calculated and displayed, resulting in a display with 12 mean frames (84-250 msec., Fig. 1 - 27b). The process is to initially select a relatively large epoch, (0-500 msec.) and by inspecting the 12 mean maps, decide that the occipital positivity deserves scrutiny, then concentrate upon a smaller epoch of interest, from 84-250 msec. Obvious time-savings can be achieved by judicious use of such an approach.

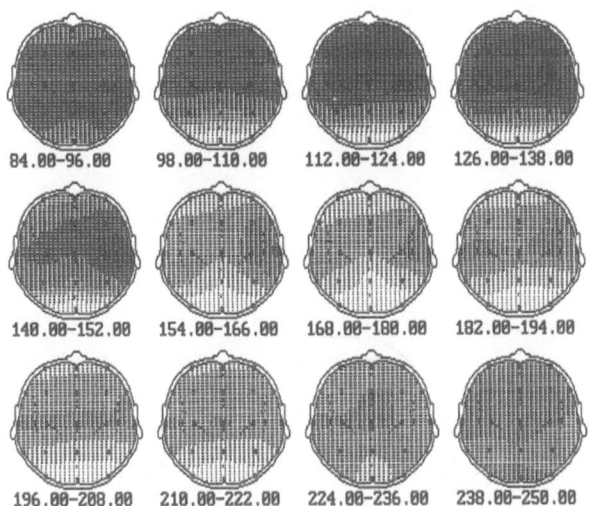

1-27b: Summary maps: expanded time scale of Fig. 1 - 27a, allowing more temporal and spatial details to be seen for the interval 84 - 250 msec.

1.6 IDENTIFICATION OF TOPOGRAPHIC FEATURES

Generally, interpretation of topographic maps is done by pattern recognition. Before any intelligent interpretation can occur, it is necessary to have a good understanding of what constitutes significant patterns, and be able to differentiate those from artifacts. An important question is: what patterns does one look for?

Due to the way maps are constructed, one should start by analyzing dominant spatial features, and to discern temporal features by following a series of maps, or by inspecting the tracings of the channels of interest. Some of the *primary* spatial features that ought to be recognized first are peak, gradient and symmetry. Next, there are *secondary* features like spread (the pattern of movement of a peak), and regions of persistent hypoactivity. Finally, one should formulate what the configuration of intracranial neuronal generators is that gave rise to this particular potential map. The objective of this last *tertiary*

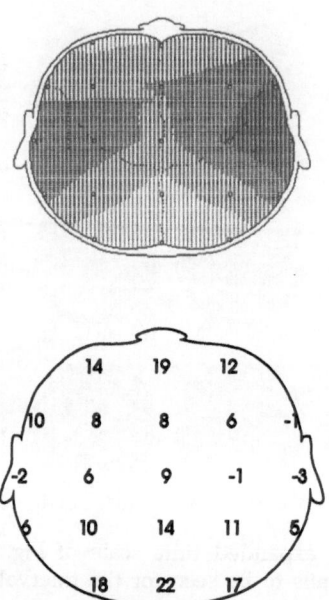

1-28: Map features. Top: map of P100 of normal PR VEP; bottom: a numeric matrix of the actual electrode voltages superimposed upon the head outline. Note the positive peak at Oz. White = +, black = -.

step is of course to be able to link particular anatomic structures to an observable electrical feature or activity. While this may not be totally successful, even partial success would provide useful insight under many circumstances.

Fig. 1 - 28 illustrates a frame from the pattern-reversal (PR) VEP of a normal adult subject. The primary features are the positive *peak* (maximal at the Oz site), which is quite *symmetric*, and whose potential field falls off with distance from the maxima (Oz). In other words, there is a posterior-anterior field *gradient*. Actually there is also a medial-lateral gradient as well.

The gradient is calculated by the difference between potential values, and can be seen from the maps by how tightly or loosely spaced adjacent colour bands are. A large gradient is characterized by many colour bands tightly spaced together. In Fig. 1 - 28, the maximum gradient is the area which forms a semicircular region

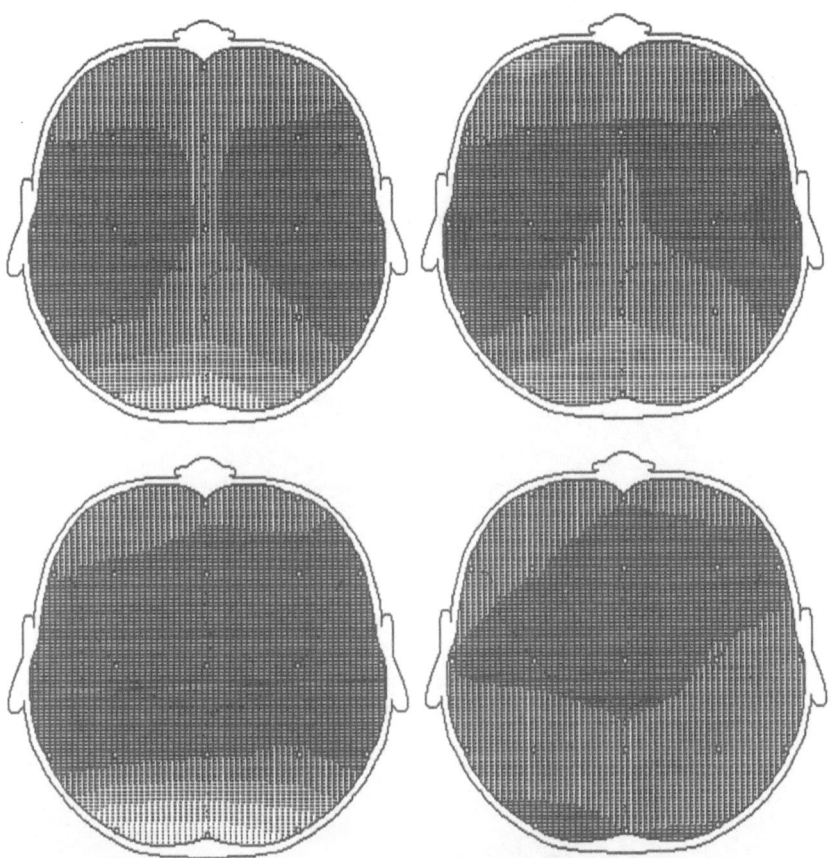

1-29:   Normal FVEP showing 4 instantaneous voltage maps. Top left: at 94 msec, start of the occipital positivity; bottom left: at 108 msec, the positivity evolved to be more focal; top right: at 170 msec, the occipital positivity decaying, and some anterior spread noted; bottom right: at 240 msec, occipital and parietal activity have disappeared.

around Oz. The gradient right at Oz is in fact small, as it is a local potential peak. Far away from Oz, the gradient drops off, and in fact is flat in the anterior head.

If one follows these spatial features over time, say from 90 to 240 msec. (Fig. 1 - 29), one can clearly observe the birth and decay,

1-30: Spike evolution. Map series of the focal discharge followed over time, from top to bottom, starting on the left column. The initial negativity appears at C4, spreads to T4, then T6, before decaying, switching to the positive phase at C4 (top right), which becomes more diffuse before decaying in turn. White = +, black = -.

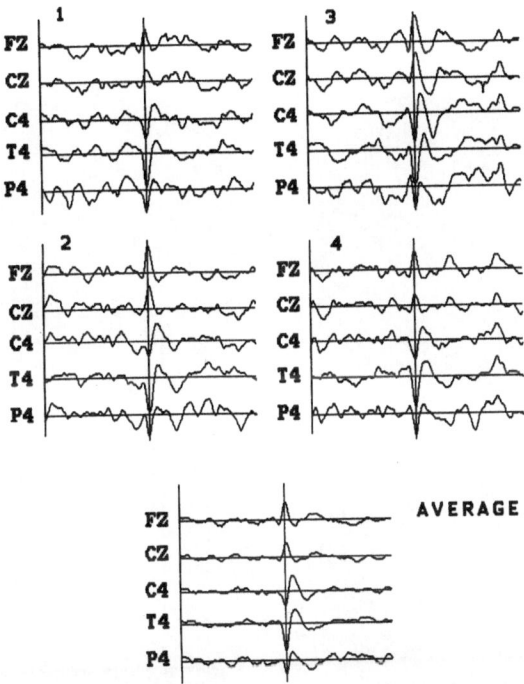

1-31: Independent rolandic spikes and their average. The top 4 tracings are individual spikes from the same patient; the bottom is the time-aligned average of these and additional different spikes from the same patient, showing an increased signal to noise ratio.

and anterior movement of the occipital positive peak. The observed movement has to be further analyzed: is it due to volume conduction effect? Or is it attributable to actual neuronal propagation?

Volume conduction is instantaneous, and its topographic signature can be likened to a mountain arising out of the sea. There is one peak amplitude point, which is at a fixed location. Surrounding it are regions of lower amplitude, whose value falls off with increasing distance from the peak. The whole potential field will grow or decay (rise and fall) together with time (i.e., *synchronous*). Contrary to this, propagated signals show a gradual progression

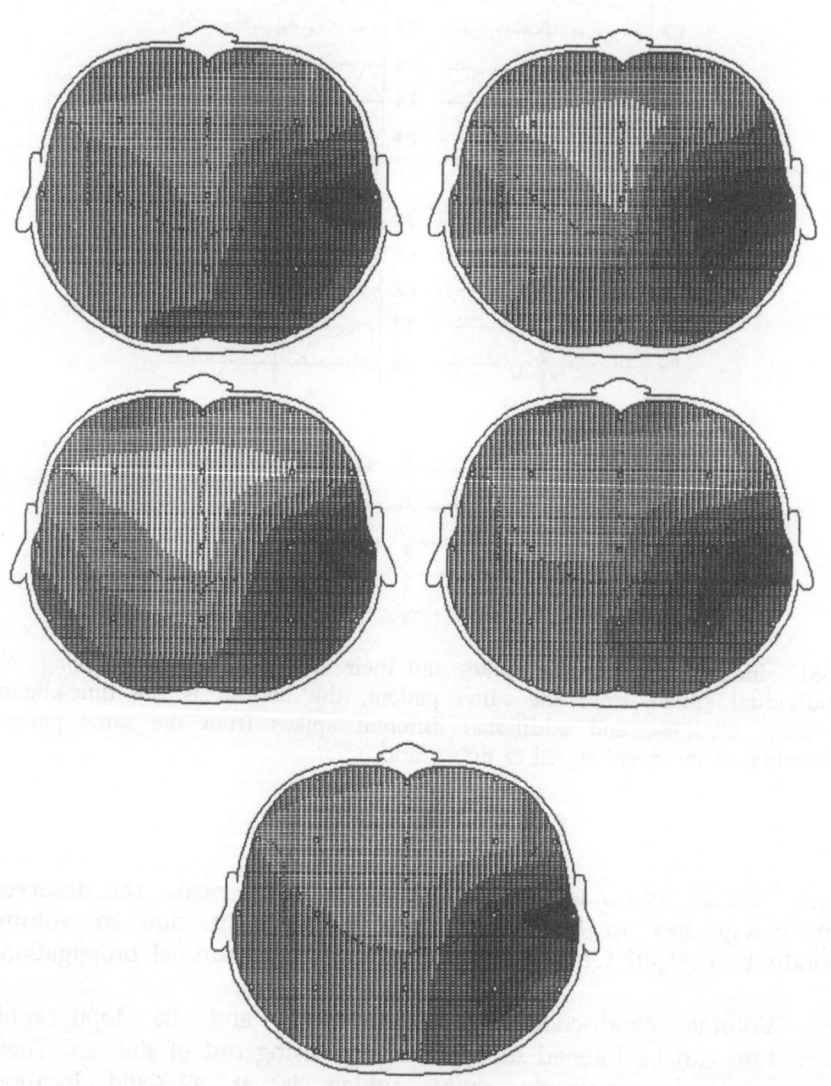

1-32:   Spike maps. Same data and arrangement as Fig. 1 - 31, showing the corresponding maps at the cursor (spike apex). White = +, black = -.

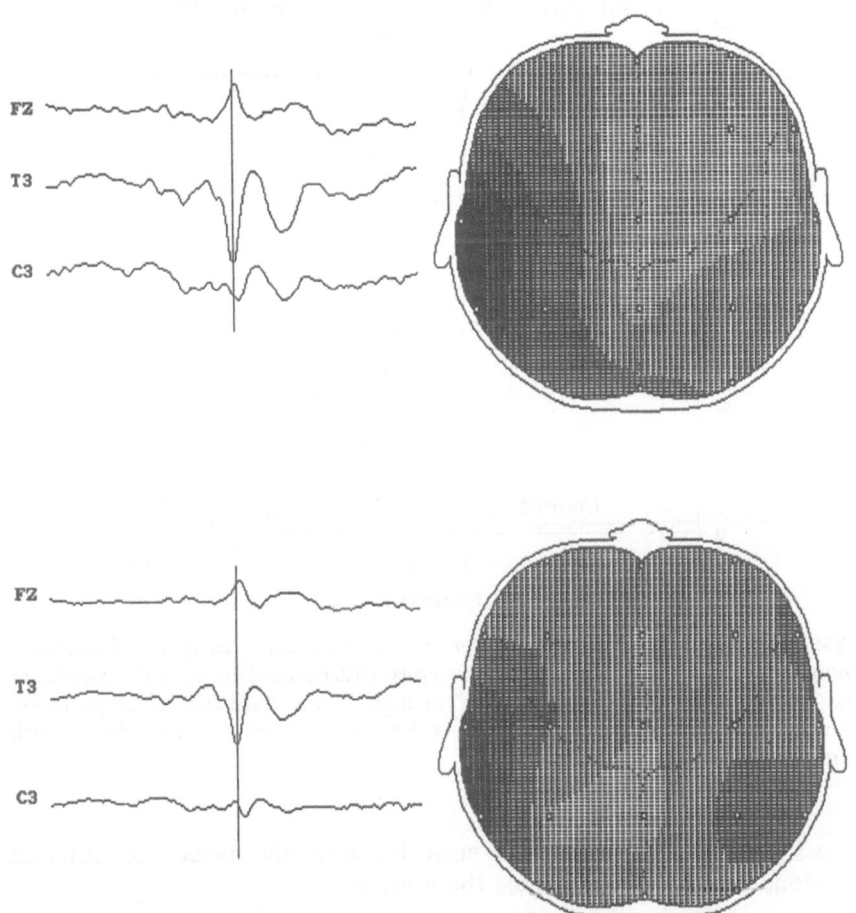

1-33: Temporal spike. Tracings and maps of a left temporal spike (top), and the Hjorth-derived representation (bottom).

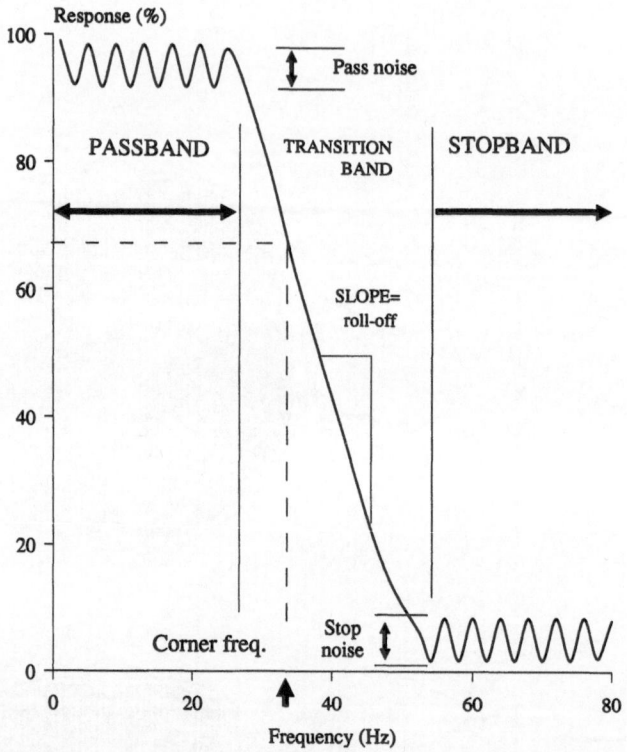

1-34:   Digital filter. Ordinate = filter output response, abscissa = frequency. There is usually some degree of uncertainty (fluctuation) in both the passband and stopband, denoted by the "noise" swings. In the transition band centering around the corner frequency, the slope is determined by the roll-off parameter.

across space, with time lags seen between the peaks at different locations. Fig. 1 - 30 illustrates these cases.

Despite having identified an example of propagated peak movement, it is usually not possible to state clearly what is responsible for this movement. There are several possibilities: real movement of a discrete neuronal generator; in-situ rotation of a discrete generator; complex interaction of 2 or more generators; complex pattern of membrane potential changes from a large (non-discrete) or extensive neuronal layer.   These are difficult situations,

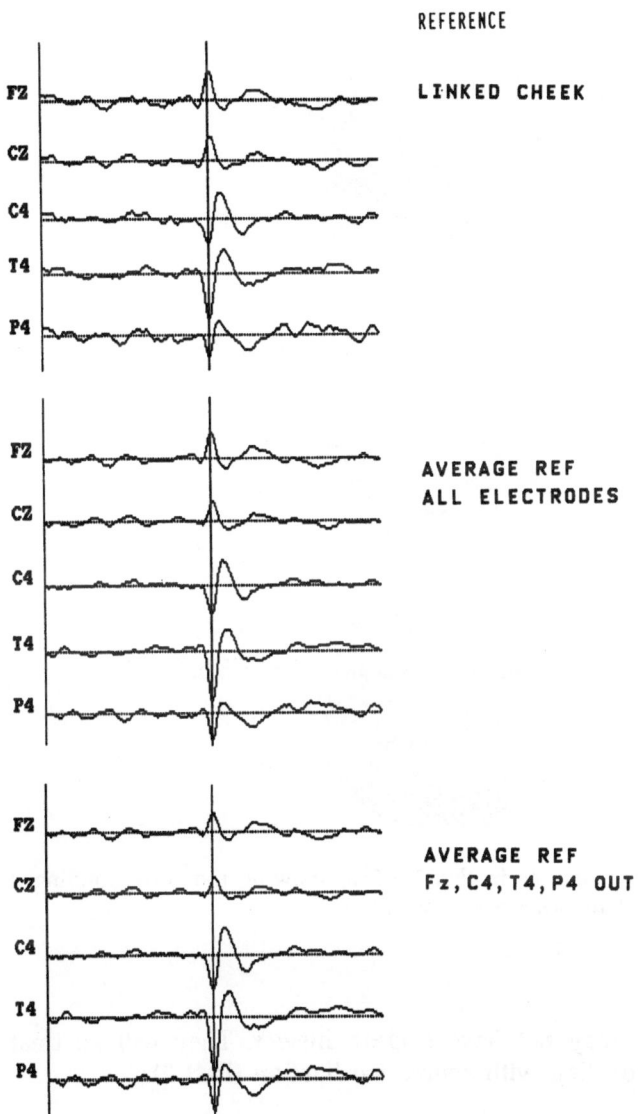

1-35a: Average reference. Example of a rolandic spike segment, displayed with the original referential derivation (top), averaged reference using all electrodes (middle), and a selective average (bottom). Although no real differences are noted here, subtle changes have been introduced in the waveform morphology due to numerical rounding-off.

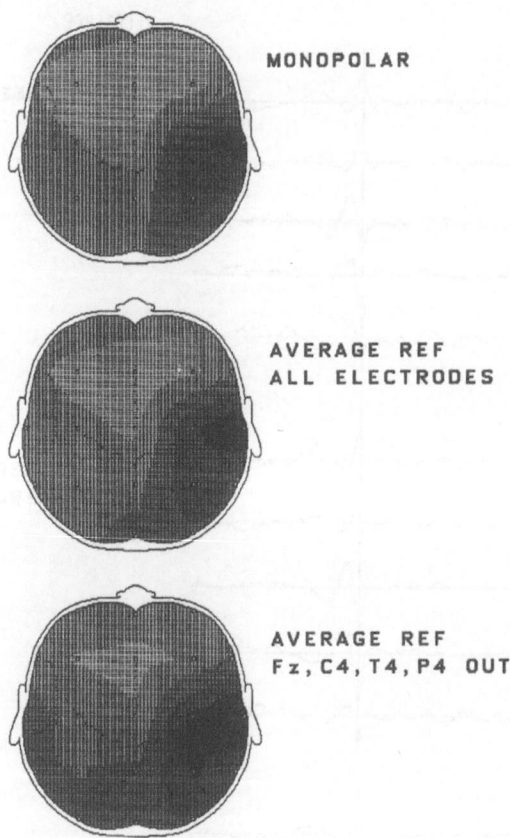

MONOPOLAR

AVERAGE REF
ALL ELECTRODES

AVERAGE REF
Fz, C4, T4, P4 OUT

1-35b: Same data as in Fig. 1 - 35a, showing the corresponding maps at the spike apex (cursor). Black = negative.

and some may not have a clear answer. They will be treated in the chapters dealing with source localization (Part 2).

Upon inspection of such map series, it may be possible to discern regions of amplitude hypoactivity or hyperactivity. This kind of *regional deviation*, best shown by unsigned integral maps, may reflect underlying abnormality. Due to the uncertainties of intracranial localization using only scalp data, the abnormality is not necessary directly underneath the deviant region.

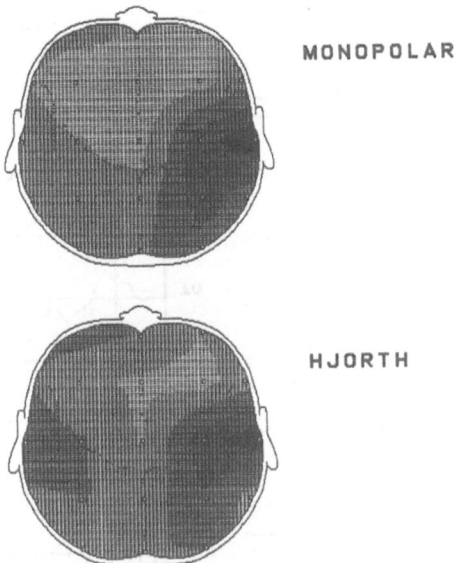

MONOPOLAR

HJORTH

1-36: Source derivation. Same data as in Fig. 1 - 35.

To recapitulate, the process of visual interpretation can be broken down into 3 levels. The *primary* features are peaks, symmetry and gradient. The *secondary* features are spread and regional deviation. At the *tertiary* level, the significance of all the known features is synthesized into a working hypothesis concerning the location of the abnormality and its extent.

1.7 SPIKE MAPPING

The procedure for spatial analysis of epileptiform transients in routine EEG can be depicted in the following manner. Several spike examples are inspected so that the degree of reproducibility can be estimated. Based on the potential field at the spike apex, a mental topography is conjured up. A statement can then be made as to the location and spread of the transient, with some conjecture as to the neuronal substrate involved.

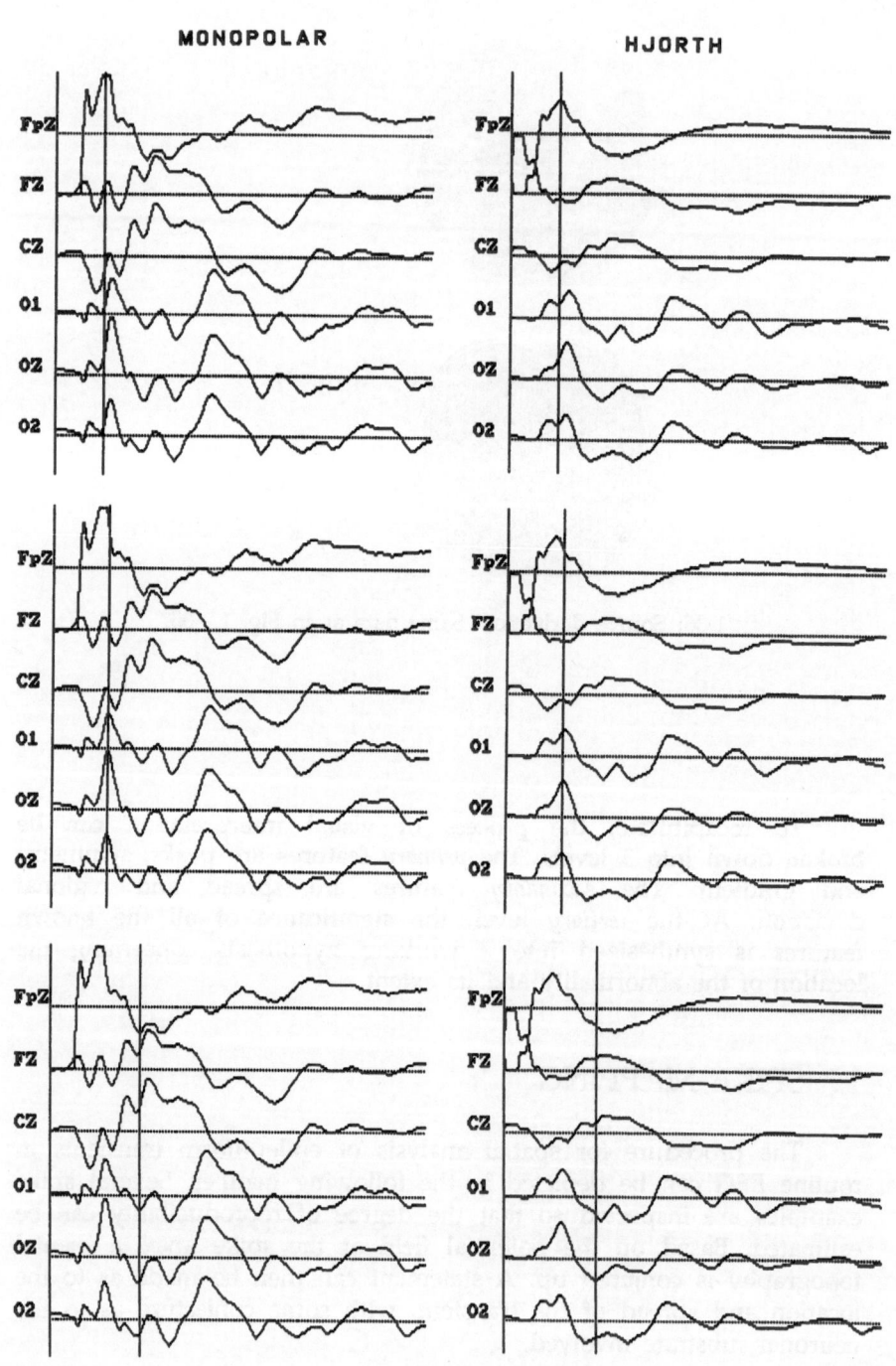

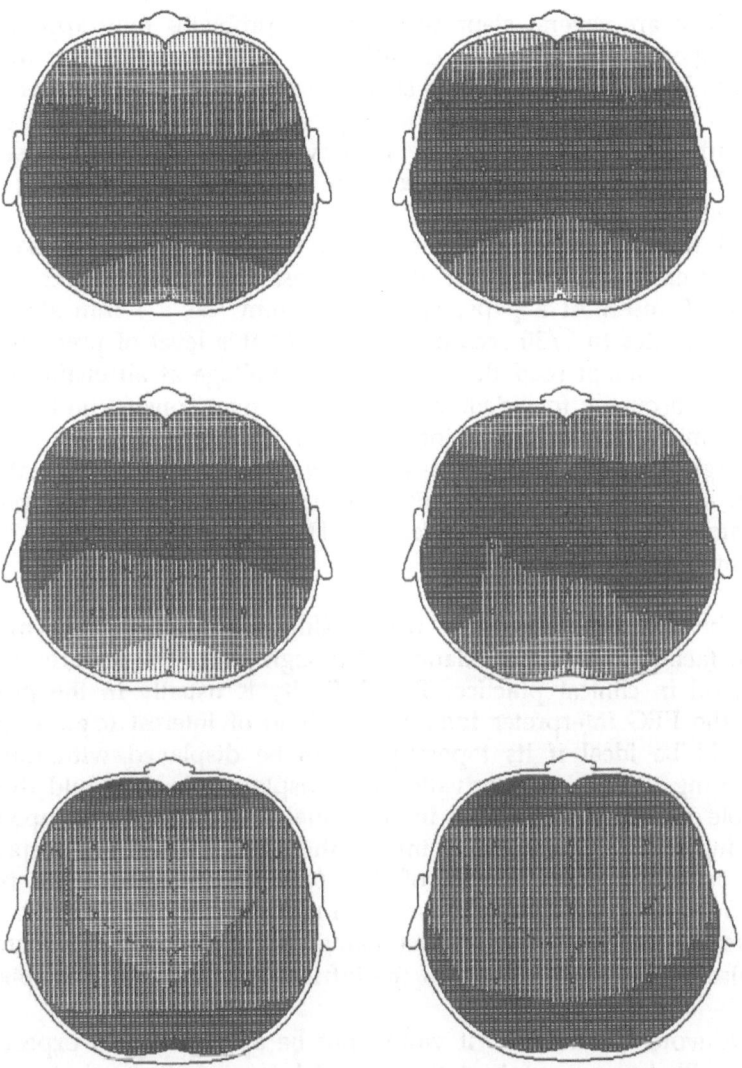

1-37b: Same data and arrangement as Fig. 1 - 37a, showing the corresponding maps at the cursor. The top left map has marked frontal positivity (white), being confined to the Fp1 and Fp2 electrodes in the Hjorth maps.

1-37a: Reference contamination. VEP with eye artifact, tracings at several time points before & after the artifact.

There are several steps where major problems may appear. The most important is that time resolution for ordinary pen-paper tracings is poor. We can estimate that such error may be in the order of at least 15 to 30 msec., which is much greater than the time needed for a neuronally conducted signal to traverse the distance of several inter-electrode distances. This estimate is based on studies of fiber size distribution in man (Nunez 1981), suggesting that the large intra-cortical fibers conduct in the order of 6-9 m/sec. Further, Bancaud (1974) measured conduction delay across the frontal lobes to be around 10 msec. At a paper speed of 30 mm/sec, a 1 mm alignment error translates to 1/30 sec., or 33 msec. At this level of precision, the human eye cannot read the instantaneous voltage at all channels to a sufficient accuracy for adequate isopotential map construction so as to obtain map to map precision of 5 or 10 msec. It then can be meaningless to spend the time and energy to produce a mental map which may in fact be erroneous. Furthermore, such a mental process is tedious, and relatively subjective. There may not be good intra- and inter-observer consistency.

There is little argument that a simple and convenient method which facilitates the appreciation of topographic EEG features would be useful in clinical practice. The difficulty is usually in the process. Once the EEG interpreter identifies an item of interest (e.g., a spike), it would be ideal if its topography can be displayed with minimal fuss, using the optimal derivation and display scale. It would then be possible to scan the tracings in the usual fashion, and call upon the mapping facility at various points of the data, whether for inspection only, or to store the segment of data (e.g., spike segment) for future use. Such a process would be an improvement upon the current practice as described in the first paragraph. These selected segments can also provide an estimate of the intrinsic variability of the spike.

A word of caution: it would not be appropriate to expect that there will be powerful statistics which can indicate immediately whether a given segment of EEG is "normal" or "outside normal limits". Although this scenario is possible, it can only be justified if a thorough understanding of the data is available, or it has been fully mathematically characterized somehow. Then appropriate statistical tests or rules can be applied to the data in question. Implicit is the assumption that the data obeys certain rules expected by the

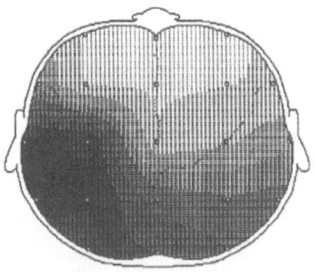

MONOPOLAR

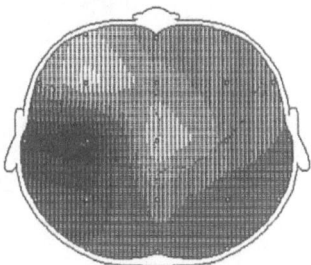

HJORTH

1-38:   Twin foci from spike map data: the original map shows a broad left temporal-parietal negativity (black) and bifrontal positivity, while the Hjorth map shows the positivity clearly having 2 foci, at the F3 and Cz locations, while the negativity is more confined to the T3 - C3 area.

statistical tests, whether the data must have Gaussian normality, without correlation between data points, or of equal variance etc. EEG data seldom fulfil these requirements, and thus many of the parametric test results (t-test, z-test, F-test etc.) should be interpreted with a great deal of care. Such issues are detailed in Part 4.

There are several ways whereby the process of mapping a spike might be improved. Fig. 1 - 31 shows an example of a rolandic spike present in the scalp EEG of a patient. Each of the 4 tracings contain an independent spike, aligned so that the negative spike apex is at the center of the short (1.28 sec.) segment of data, which is indicated

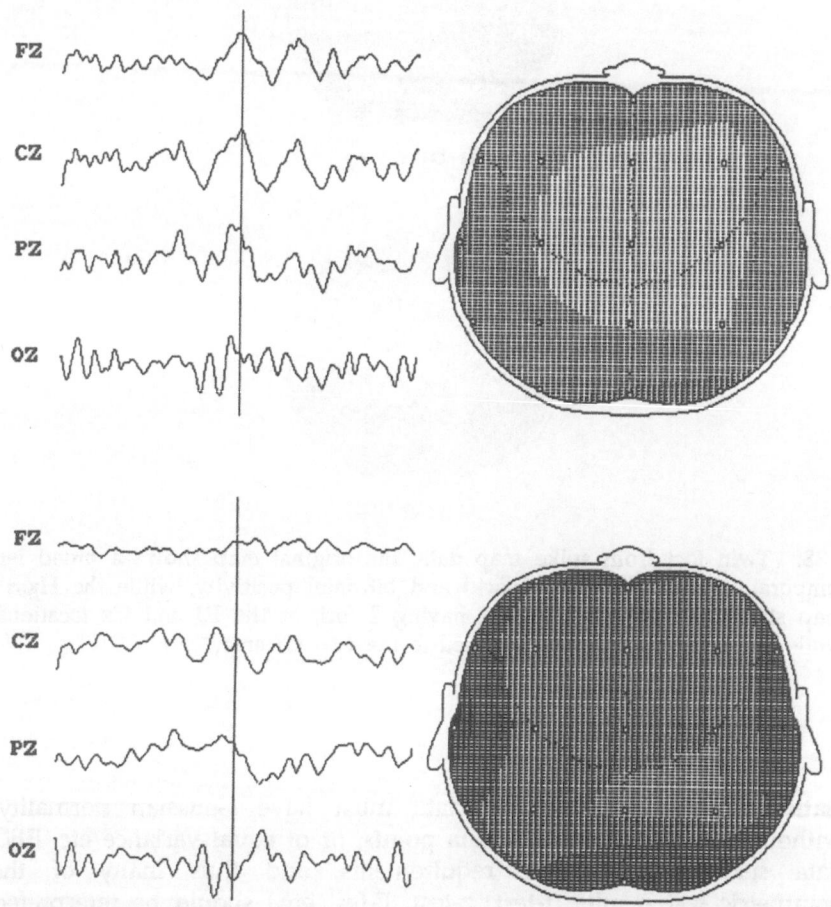

1-39:   Diffuse delta: example of poor result with source derivation. The original data before (top) and after source derivation (bottom). The strong vertex delta activity has been almost completely attenuated. The cursor was kept stationary (white = +).

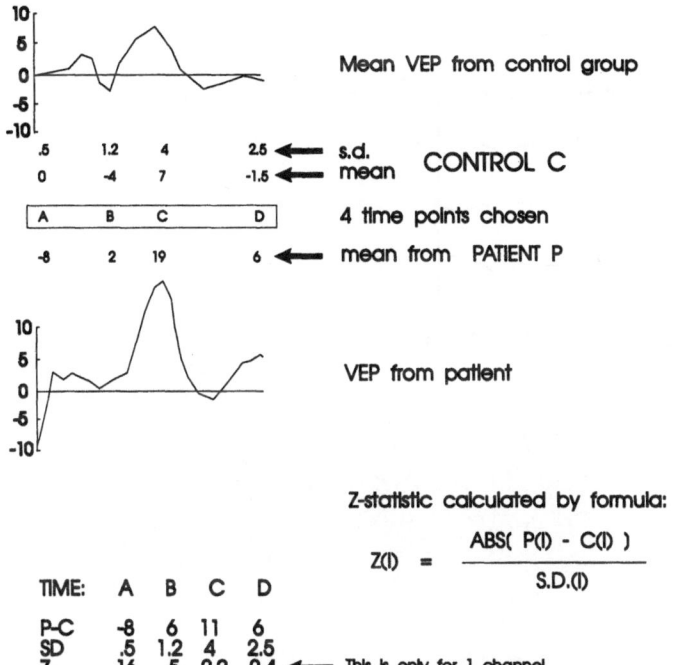

1-40: Z-statistic calculations illustrating only 1 channel for brevity. The mean and standard deviation at each data location for all time points are calculated (top), for both the control group and the individual patient. Z values at 4 time points are calculated here.

by the cursor. The corresponding maps in Fig. 1 - 32 have been constructed from that time point. It can be seen that there are 2 clear polarities, with the negative peak centered over right mid-temporal (T4) and the positive peak centered over mid-frontal. In these examples from the same patient, it is clear that while the peak activity consistently involves T4, the rest of the field has a good deal of variability. The source of this variability is the background EEG signals (alpha, sleep potentials, eye blinks etc.) which are random in nature in relation to the spike itself. They cause electrical interference or "noise", and can be minimized if the spike activity is of high amplitude while the background is comparatively of low amplitude: an instance where the "signal to noise ratio" or SNR is high. This is

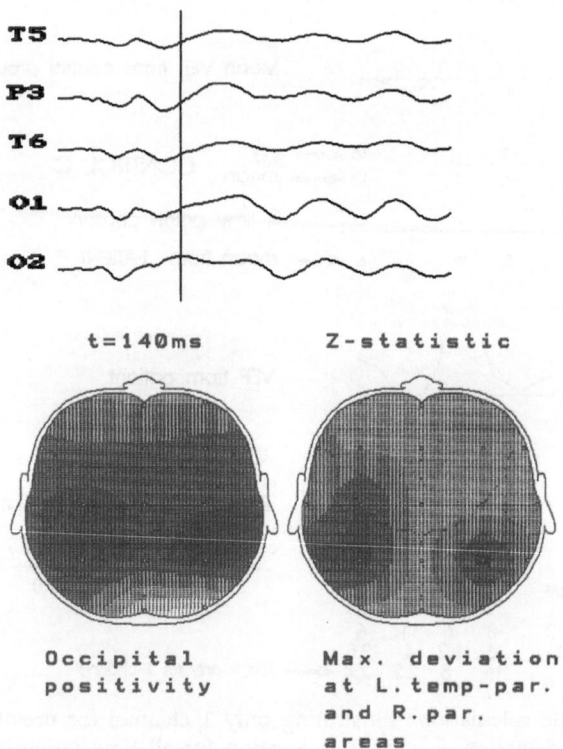

1-41: Z-map. Top: VEP tracing showing depressed left occipital response, seen on the map (bottom left) at 140 msec. as an asymmetry. The z-statistic map (bottom right) calculated by comparison with control shows maximal deviations (black) at T5, P3 and P4 electrodes.

seldom the case in practice. This noise is present as voltage fluctuations in the tracings just before the onset of the spike itself. The averaged tracing at the bottom shows a marked decrease in this pre-spike noise, and a cleaning-up of the spike component.

By a process of averaging, several spikes from the same focus may be combined to yield an averaged spike which is of higher SNR than each of the individual spikes. The mathematical and physiological basis is the same as that of evoked potential averaging, and the same assumptions and limitations also hold. It has to be

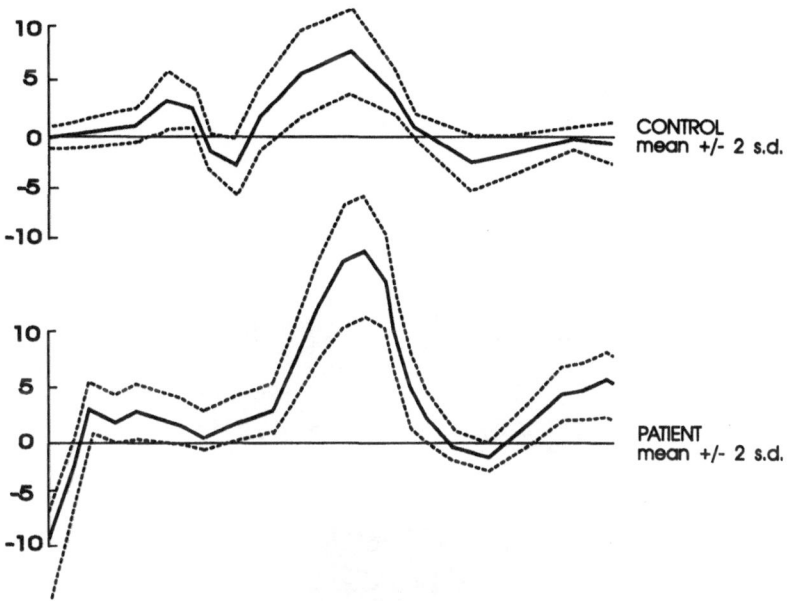

1-42:   t-map construction from computation of the means and standard deviations of data from the 2 groups in similar manner as Fig. 1 - 41.

ensured that only similar spikes are averaged together, and not spikes from independent foci. Prior to mathematical averaging, individual spikes must be time-aligned using the negative peaks at the same channel. Fig. 1 - 32 shows the averaged spike from the data in Fig. 1 - 31. The tracing now shows a cleaner appearance due to the cancellation of the random background signals, and the spike morphology is much more sharply defined. The maps at various time points likewise have smoother contour outlines.

Fig. 1 - 33 shows an example of focal spike from the temporal region together with the source derivation map. Inspection of the tracings (bipolar montage) did not easily lend itself to a mental imagery of the topographic field. By using reformatting of montage, other tracings of the exact data segment can be obtained.

In the case of generalized 3 per sec. spike and wave, the difficulty with mapping is the relatively small amplitude spike component preceding the large wave component.   In an attempt to

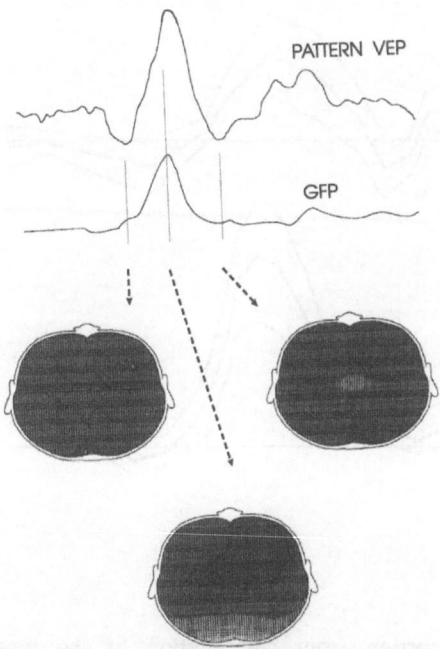

1-43:   GFP calculation. Top: tracings of normal PR VEP recorded at Oz and its GFP. Bottom: 3 time points are chosen for mapping; the one corresponding to the maximal GFP has the most hilliness.

analyze both these features, one may be tempted to use a higher setting for the high-pass filter (say 5 Hz instead of 0.1 Hz), so that the slow wave is attenuated, while the fast spike is relatively untouched. If one were to carefully inspect the data, some important but subtle differences can be found. The peak location of the spike would be unstable, and clearly influenced by the filter settings used. This is an example of the well-known but often ignored effect of phase shift or phase distortion which was brought about by tight filtering. Such topographic instability may not be obvious by visually inspecting the tracings.

A solution to this lies in the use of zero phase-shift digital filtering discussed in the next chapter. For now, such a technique can result in attenuation of the slow wave components, while accentuating the spike components, all without introducing any phase shift.

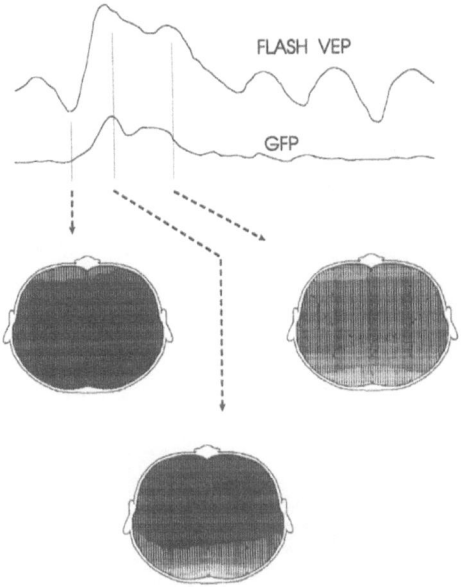

1-44:   Flash VEP. Same format as 1 - 43; note the more complicated GFP peak morphology as compared to the previous example.

1.8 POST-PROCESSING

The ability to process the data in a myriad of ways has always offered great potential to EEG/EP analysis. However, post-processing is not new as examples can be found in the literature dealing with many analytical techniques ranging from simple to complex (Gevins 1987). The results of such data manipulation offer additional insight into its structure or information content, or significantly improve its SNR or comprehensibility by inspection. In order to be functional, such manipulation must be non-obtrusive and simple to use. This chapter will be limited to surveying several common or interesting approaches: digital frequency filtering; spatial filtering; reference manipulation; statistical mapping; global field power; and source localization/characterization. Frequency analysis will be covered in the next chapter.

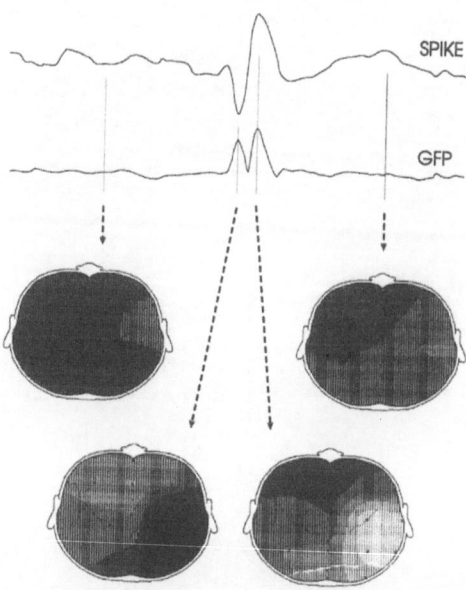

1-45:   Spike example. Top: tracing of a central spike (T4 position) and the corresponding global field power (GFP). Bottom: sequence of 4 maps corresponding to the indicated times. The bottom-most 2 maps correspond to the sharp component of the spike and have the most "hilly" topography, showing a right temporal prominence.

Analog front-end

A practical and desirable hardware system for the acquisition of EEG signals for post-processing can based on a fixed gain amplifier, properly matched to the ADC device. An anti-alias analog filter stage must be interspersed in between to ensure proper conversion. Assuming an ADC precision of 10 bits, such a front-end subsystem will allow 1,024 distinct amplitude levels, or +/- 512 bipolar levels. This can yield an input signal amplitude range of +/- 512 uV, with a step size of 1 uV. As most EEG amplifiers have a noise level of approximately 1 uV, and most EEG features of interest is less than 512 uV, this arrangement avoids the problem of having to manually match the gain to the input signal amplitude, increasing it for small spikes, or decreasing it for large delta waves. If the input range is

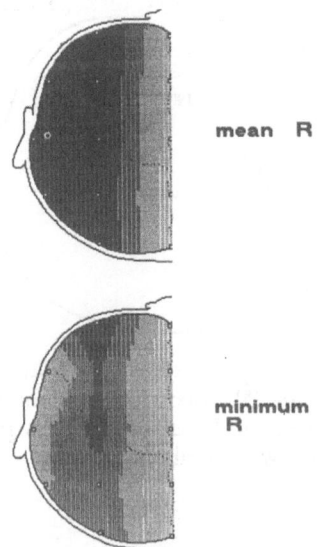

mean R

minimum
R

1-46: Correlation map. Pearson's correlation of homologous channels in flash VEP in normals. Top: average r values (R) in a control group. Bottom: R - 2 s.d. threshold map (2 s.d. below mean). Black = high correlation, white = low.

lower, and such matching adjustments are not made, then the acquired data will either clip off the tops of large amplitude waveforms (saturation), or have too coarse a precision to display low voltage details. Even better performance is an ADC precision of 12 bits, which then allows an input range of +/- 2 mV at a noise level of 1 uV.

Digital filtering

The action of analog filters can be closely simulated by digital algorithms, as can be formally shown by signal processing theory (Hamming 1983). In fact, there are published software packages available for implementing this function in microcomputers using

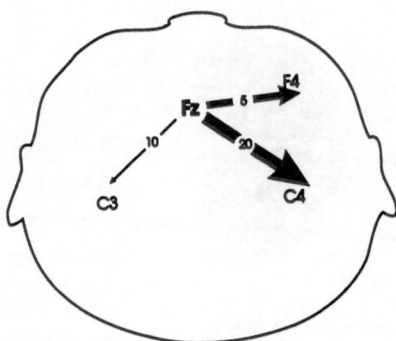

1-47:   Correlation display. Map to show correlation between 4 electrodes. Here Fz leads C4 strongly by 20 msec., leads F4 moderately by 5 msec., and leads C3 weakly by 10 msec. Thickness of arrow is proportional to the value of the correlation coefficient, while embedded value = lag time.

high-level language (IEEE 1979). It merely requires proper acquisition of raw data into digital format, and knowledge of the sampling rate. All the common analog filters can be duplicated: low-pass, high-pass, band-pass, notch, etc. Discussion here will be on the generic filter, and the reader is referred elsewhere for further details (Hamming 1983; Oppenheim and Schafer 1975; Rabiner and Gold 1975).

In practical filter design, it is important to realize that 2 broad classes can be encountered, based on whether the phase dependence with frequency is linear or not. The former will ensure smooth and predictable phase shifts, which can then be corrected by using a second stage to produce a correcting phase shift of same magnitude but opposite in direction (polarity). With non-linear phase characteristics, however, this is not possible.

Considering a low-pass digital filter, these are the important performance characteristics: corner frequency (below which there is little or no attenuation), roll-off (the rate at which attenuation occurs, or the steepness of the frequency response curve), gain (usually unity), phase change (usually dependant on frequency, or may be zero phase-shift), noise level (digital corruption of the clean input signal). In its design, the choice of algorithm is first made: Butterworth, Chebyshev, etc., based on the desired behaviour. Then the other characteristics are used to   determine the length of the filter

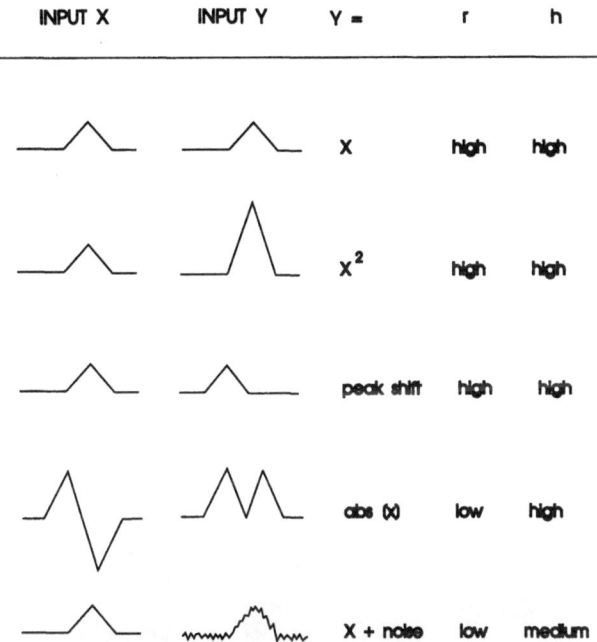

INPUT X	INPUT Y	Y =	r	h
		X	high	high
		x^2	high	high
		peak shift	high	high
		abs (X)	low	high
		X + noise	low	medium

1-48: Linear and non-linear correlation between 2 signal inputs (X and Y). The third column shows how Y has been derived from X. Comparison of r and h under different theoretical situations. The presence of random noise greatly affects the accuracy of r.

(the number of stages). There is always a compromise required in order to achieve a workable filter, one with the lowest number of stages preferably. A design with zero noise or flat response plateau simply cannot be achieved. Digital noise is seen in the form of fluctuations of the response curve. See Fig. 1 - 34 for an illustration.

Reference manipulation

If the original data had been acquired properly (ADC rate, dynamic range, gain linearity etc.), it contains all the information necessary for reformatting a new dataset as if another reference had been used. Of course the new reference must be one of the channels which had been recorded. Two specific situations, average reference and source (Hjorth) derivation, deserve mentioning.

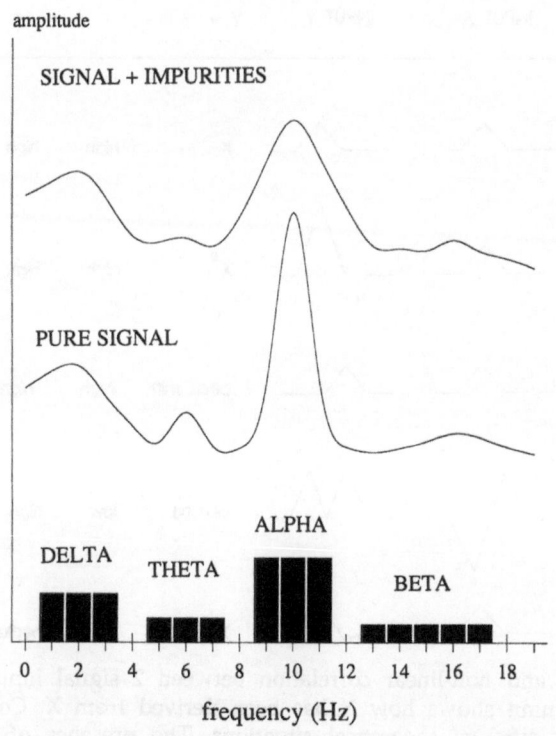

amplitude

SIGNAL + IMPURITIES

PURE SIGNAL

ALPHA

DELTA

THETA

BETA

0   2   4   6   8   10   12   14   16   18

frequency (Hz)

1-49: FFT. The input signal (top tracing) consists of the pure frequency components (or signal; middle tracing) and an unknown amount of random noise, all admixed together. The computed FFT components of the input can be seen as a bar diagram along the frequency axis (bottom).

As illustration, Fig. 1 - 35 shows a spike segment derived from the original recording using linked cheek reference, and after transformation to average reference. The most obvious effect is that the map colouration has changed, but on closer inspection, it is clear that the contour line pattern has remained constant. This is due to the fact that the transformation merely substituted another voltage value (the average of all channels at that time point) for the original (arbitrarily zero) value of the reference electrode. So the effect is the same as if the colour scale has been re-assigned. The advantage of this transformation is that the "reference" value is independent of any individual electrode. The colour bins are used to the best advantage, since the "zero" voltage colour correspond to the mid point of the scale and is defined as the average voltage (reference).

There are also disadvantages which are less obvious. Any contamination or artifact present in any electrode influence the final reference value, and steps taken to remove such electrodes from the average only serve to accentuate the second problem, namely that a true average reference should represent the entire head. In practice only the scalp is sampled, and usually with inadequate electrode density, so that at least half of the brain's surface is unrepresented. Added to this is the fact that much of the cortical surface remains within sulci and therefore "hidden" from scalp recording sites. It is not surprising that, like all reference systems, this is not ideal.

A similar approach makes use of the fact that the field gradient has information of interest. The same data in Fig. 1 - 35 is redisplayed in Fig. 1 - 36 after source derivation. As field gradient is emphasized here, the voltage peaks appear much more "sharpened" or focal. Conversely, diffuse effects are attenuated, such as reference contamination or baseline offset.

Fig. 1 - 37 shows a VEP where there is reference contamination from the eyes, showing up in the map as a diffuse peak maximum bifrontally, with a gradient going lower posteriorly. If just a single frame of the map is inspected, it would not be possible to unambiguously identify such an artifact, which may be partly due to the electroretinogram. After source derivation, these spatial peaks are limited to the 2 electrodes nearest the eyes, and the rest of the head is relatively free of activity. At a later frame, the occipital activity emerges in a much clearer fashion.

Focal events (e.g., spikes) can be more sharply defined by source derivation, as seen in Fig. 1 - 38, where 2 clear focalities have been brought out from a broad spike. Such a topography suggests 2 separate sources, as the double peaks cannot be generated by one. However, diffuse activity suffers attenuation, as seen in the delta activity in Fig. 1 - 39.

In many instances, there is an obvious clarity of spatial features brought out by the source derived maps. If the idiosyncrasies of the method is understood, the result is easier interpretation of the localization of the original data. Blind application of this method is

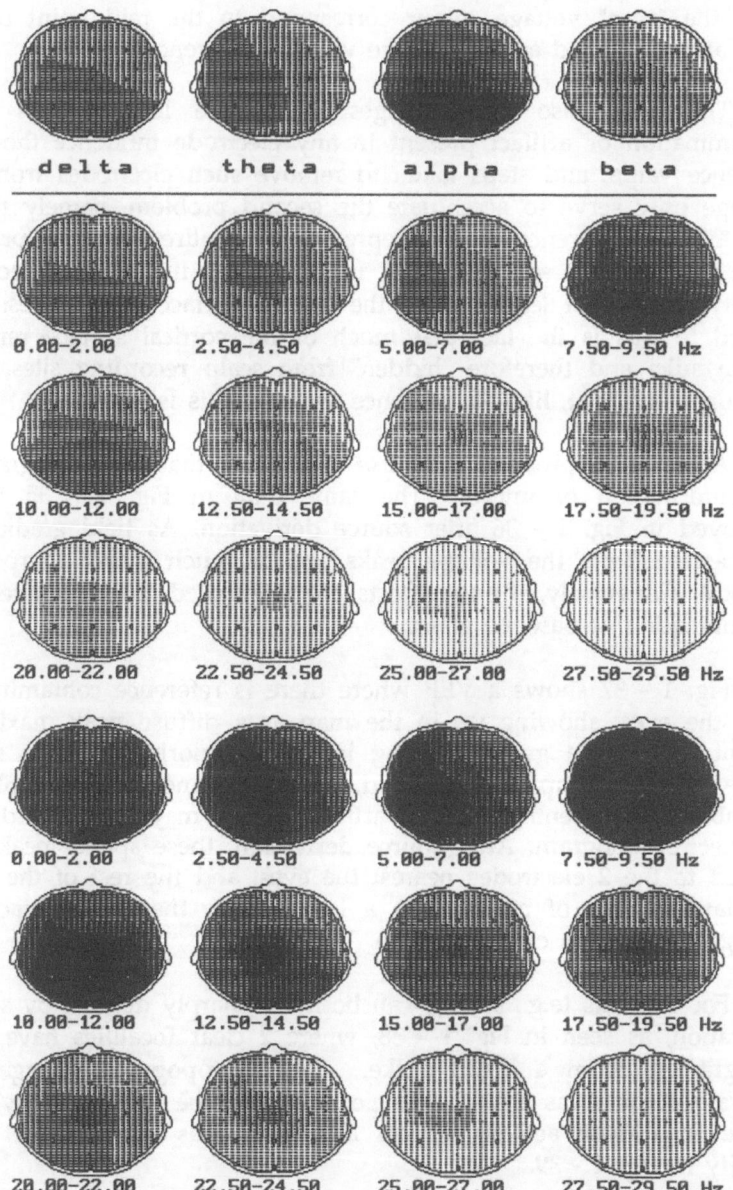

1-50: Normal FFT. Top row: frequency band maps; the next 3 rows show individual maps representing 2 Hz bins; bottom 3 rows: same data re-displayed at doubled gain. Black=high activity, white=low activity. Reference: linked cheeks.

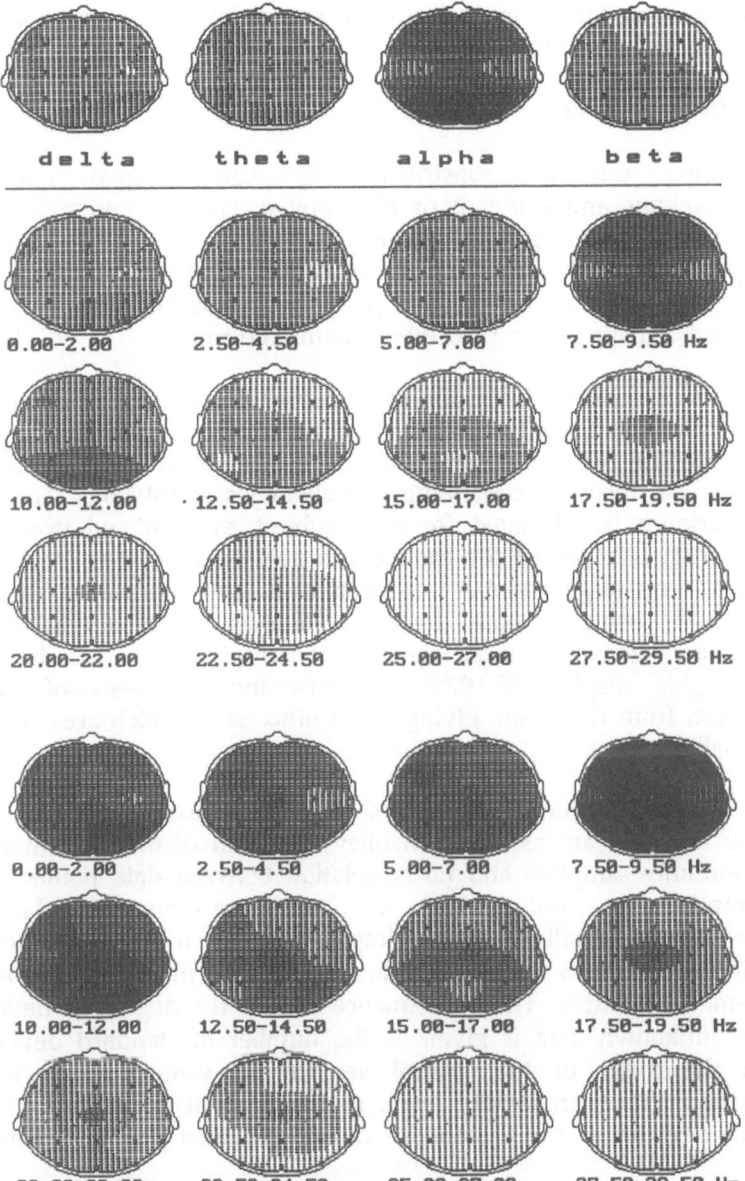

1-51:   Normal FFT. Same data and arrangement as Fig. 1 - 50 but digitally reformatted using the average reference. Note the diminished occipital predominance and the more prominent frontal component of the alpha map (7.5-9.5 Hz).

dangerous, as for example the edge electrodes of the scalp array is less accurately transformed.

Statistical mapping

Maps can be constructed by using signal amplitude (instantaneous amplitude maps of a spike), spectral parameters (FFT bands), or derived statistical values (correlation coefficients, Student's t, z transform etc.). More complex statistical results can be mapped of course, but for purposes of this part, the discussion shall be limited to t- and z-maps as the principle remains the same.

Let us consider 2 groups of subjects (A and B, representing say patient and control). It is desired to determine if they are basically different. The calculation procedure is the same for all electrodes, so that we can focus on describing one channel's calculations. The mean and variance is calculated for each subject, at each and every time point. This is repeated for each electrode's data, so that we end up with a very large number of mean and variance values for both groups. By using standard formulae, one arrives at a dataset of z values arranged as frames, replacing the original recorded data. The topographic display of these z maps indicate areas of greater deviation from the mean, giving some hints as to which areas are less "normal".

It must be stated again that so far many assumptions have been made: the data are assumed to obey Gaussian distribution (normal), be randomly sampled, and no correlation between data points. These assumptions are not usually met, and are sometimes blatantly ignored. It is usually therefore impossible for an accurate statistical significance level to be assigned, and the results must then be treated as semi-quantitative. The consequence is that the degree of deviation of the unknown data is given as the number of standard deviations from the mean of the control set (the z value), which is not translatable to a probability level. In this case, it is inappropriate to simply say that if $z = 2$ standard deviations, then p is less than 0.05 etc.

There are some situations and conditions which permit a clearer statistical statement to be made. In the case of Student's t statistic, the calculation compares replicated trials of a given subject under 2 different conditions (e.g., before and after treatment). Essentially the same mean and variance computations are done, and at the end we see if the two means differ significantly after accounting for the respective variances. If the degree of freedom is accurately known (not usually possible for the same reasons as stated above), then a probability significance level for the difference can be stated.

Fig. 1 - 40 to 1 - 42 illustrate how t- and z-maps are constructed, and includes an example.

Global field power

In the quest for spatial features which may be of help in characterising maps, the degree of "hilliness" of the potential landscape was thought to be of use. A measure called "global field power" or GFP (Skrandies and Lehmann 1984) provides an objective indication of this, and can distinguish maps that are lacking in potential peaks (i.e., a "flat" field) from those that have significant peaks. It is calculated by the sum squared deviation from the mean of every recorded (i.e., electrode) point of a given map (i.e., a frame). For a flat field, there is little deviation from the mean, as every point has a similar voltage value. For a field with one or more large amplitude peaks, there will be a greater total deviation. Fig. 1 - 43 illustrates this.

In this example, the relatively simple PR VEP waveform generates a simple GFP tracing, with only a single peak corresponding to the P100 peak. For other more complex waveforms, the resultant GFP tracing is more complex, and sometimes difficult to interpret. Fig. 1 - 44 shows a normal FVEP with a more complex GFP. It is at times ambiguous or misleading to rely on the maximum GFP peak as guidance to the main occipital component.

Another example of the use of GFP as a data reduction tool is given in Fig. 1 - 45. A sequence of maps from a 2 second segment of EEG containing a spike discharge is depicted, with their calculated GFP tracings. There is a clear correlation of the GFP with the hilliness of the map, and the maximum GFP correspond to the spike apex. There is more than one maxima in the GFP, corresponding to the different phases of the spike waveform. It is useful to display and view the map frames just before and after the peaks in the GFP tracing, thus effecting some degree of data reduction.

Correlation analysis

Gevins (1987) used simple cross-correlation to study the relationships between scalp regions during cognitive tasks. He demonstrated definite patterns of high correlation associated with specific visual-motor tasks (Gevins 1989). Mars and Lopes da Silva (1983 and 1987) had previously described other methods of calculating association between channels, based on average amount of mutual information. Finally Pijn et al. (1989) described a simplified non-linear correlation (h^2) which in cases of linear association is identical to the Pearson correlation coefficient (r^2). h^2 performs better than r^2 in data which have non-linear relationships, or in the presence of interfering noise.

The display of paired channel data presents a unique problem of display. Whereas amplitude data from each head location is associated with 1 channel, correlation values are associated with a pair of channels. Thus another dimension has to be provided for in the display map. Fig. 1 - 46 illustrates this using flash VEP in normals, with correlation performed for homologous electrode pairs. For this case involving only 8 electrodes (omitting the midline electrodes), it is possible to display one hemisphere only, with the T3 display value representing the correlation of T3-T4 for instance. The figure shows maximum correlation between occipital and posterior temporal pairs, then decreasing anteriorly.

In the event that more complex pairings of channels are to be tested, for instance if every channel is to be correlated with each of the other 19 channels, then the number of pairings increase to 342 (if n electrodes are present, then there are n^2-n pairs). The difficulty is that all this information have to be accommodated within a flat display of the head.   One solution is to use arrows to link respective channels while retaining the direction for representing polarity, and varying its thickness in proportion to the magnitude of the coefficient. In order not to overly clutter the correlation map, only the most significant values are plotted. Fig. 1 - 47 illustrates a simple example.

Turning to signals with non-linear relationships, r^2 or r may not be appropriate or accurate under some circumstances. Proposing the use of h^2 or h as an alternative in EEG signal analysis, Pijn et al. (1989) and Lopes da Silva (1989) described examples of their use. Fig. 1 - 48 gives some simple comparative examples. Let X and Y denote the input data. Under some conditions, both the linear (r) and non-linear (h) estimates of correlation give comparable results as in the first 3 cases of Y = X (a straight copy), squaring or a simple peak shift. For the 4th case of rectification (taking the absolute value), h still gives a higher value than r, despite the obvious relationship (i.e., association) of X and Y. Of special interest to us is the case where white noise is added to the data. Here r performs poorly, while h will show a somewhat higher value. In dealing with EEG data, it is very unusual for noise not to be present.

Source localization

In dealing with this complex topic, the theoretical and mathematical basis, anatomic modelling and assumptions, and other details can be found in Part 2. The brief treatment which follows is merely an introduction to allow the reader to decide whether to take on the daunting task of reading Part 2.

In order to put things in perspective, while a neuronal generator configuration is ultimately responsible for the potential map that is recorded, how to represent it in a way that is realistic and yet not overly complicated is a problem. The concept of a "current source" as a simple "battery" at a certain point in space has certain utility. It is really a mathematical device that happens to be easily understood because of its intuitive nature. It is then possible to use a combination of such sources to model many complicated real-life situations.

In simple terms, a source is merely a set of 6 numbers, each describing a particular aspect or facet: these are location, amplitude and direction. In this sense, the set of numbers form a simple and objective descriptor of the entire map, and may be used as an *index*

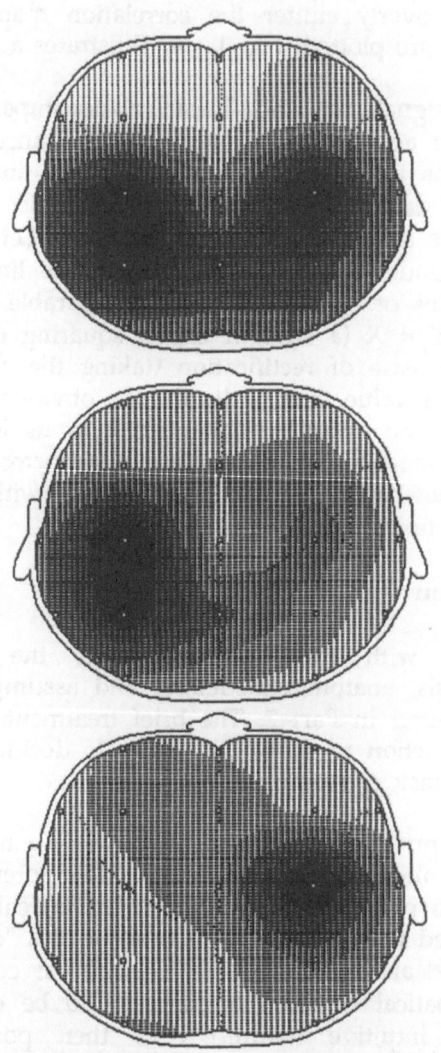

1-52: Mu rhythm. Top: spectral band maps of a normal subject under resting eyes-open conditions. Note the effect on the central mu rhythm when the contralateral hand is stimulated: left (middle) and right hand (bottom). The mu rhythm is selectively attenuated.

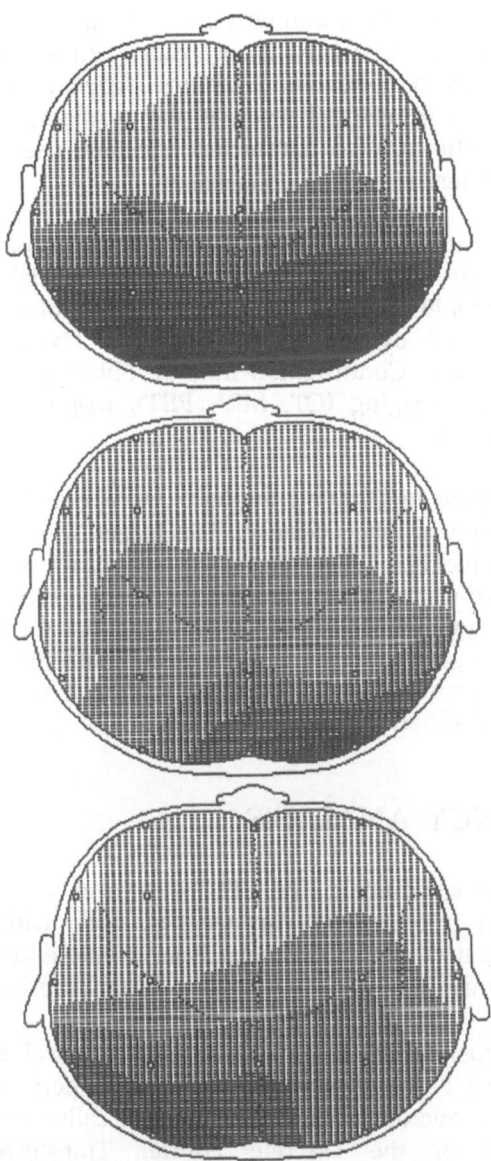

1-53: Alpha rhythm. Top: resting eyes-closed condition (top), showing strong bi-occipital alpha. In this unusual case, monocular eye opening also selectively affected the contralateral alpha rhythm: right eye open (middle), left eye open (bottom). Care was taken to rule out the presence of artifacts.

which describes the map, which was in turn produced by a particular generator configuration. In a diagnostic sense, such an index may be used to discriminate between 2 or more disease or treatment conditions, or between patient and control. Applied in this manner, location accuracy is immaterial, even if the resultant source location was extracranial. The location parameters are then used only for map description, characterization and comparison.

If the head model and mathematical analysis procedures can be made more accurate, source solutions with good spatial accuracy may be achieved. It would then be possible to contemplate the anatomic structure(s) suggested by the source solutions. Hypotheses may be formulated and tested. Confirmation may be obtained by independent means: radiologic imaging (CT, MRI, PET), magnetoencephalogram (MEG), among others.

Recently, dipole localization method (DLM) has been applied to the study of temporal lobe epilepsy. The intent was to determine if the clinical outcome can be predicted on the basis of scalp, cortical and depth recordings of interictal and ictal EEG. It was found that DLM and spike topography were both useful in this respect, with specific patterns being associated with good outcome after surgery. The preliminary results indicate that it has good potential to be a useful clinical tool (Ebersole 1989).

1.9 FREQUENCY ANALYSIS

In the frequency domain, the EEG waveform (a time series) can be broken down into several components, each with a particular frequency and spatial distribution. Analysis of these components can yield additional insight and perhaps useful diagnostic information.

The decomposition of the time series into spectral data is usually carried out by the Fourier technique. A digital algorithm suitable for fast processing by microcomputers is available, called the Fast Fourier Transform (FFT) or the Discrete Fourier Transform (DFT). Its application is shown in Fig. 1 - 49.

There is a clear relationship between the ADC sampling rate, the length of the time series (epoch in sec.), and the spectral parameters

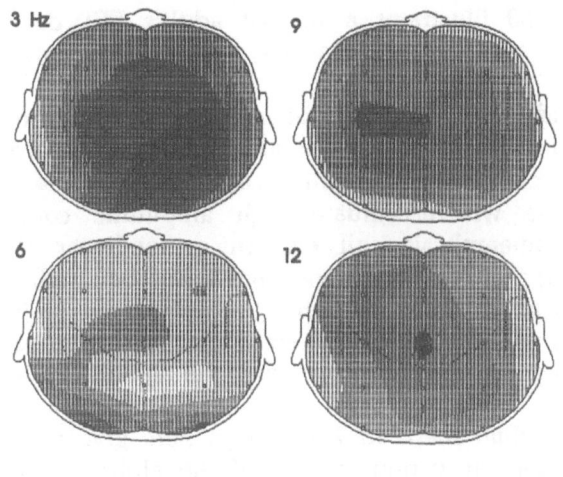

01 〰 Cz 〰

1-54: Steady-state FFT. Pattern reversal stimuli at 3 Hz: the FFT components can be seen at 3 Hz and higher harmonics. Tracings of the FFT for the O1 and Cz electrode locations are shown at the bottom. There is posterior dominance at 3 and 6 Hz, and vertex dominance at higher frequencies. White = low, black = high amplitude.

of frequency resolution (the step size between bins in Hz) and maximum frequency component (Fmax in Hz). Fmax is equal to half the ADC rate, whereas the step size is equal to the reciprocal of the epoch. Hence, if a 2 sec. epoch is used, and the ADC rate was 128 Hz, then the FFT yields spectral components starting from 0 Hz to 64 Hz, in 0.5 Hz steps. If a greater frequency resolution is desired, then a smaller step size has to be achieved by increasing the epoch to 4 sec., for a 0.25 Hz resolution. If the frequencies above beta is not of interest, then it is unnecessary to sample at greater than 64 Hz, which yields a spectrum of 0 - 32 Hz. It is important to prevent the occurrence of aliasing error, as pointed out in Chapter 1.2.

Fig. 1 - 50 illustrates a normal adult's FFT data. It may be displayed as a set of spectral tracings, with spectral power plotted as the y-axis, and frequency as the x-axis. The square root of the power (the amplitude, which is linearly scaled) can also be used, and allows a better appreciation of the relative contributions of each component. Power tracings will accentuate high amplitude components (i.e., alpha), and de-emphasize all else in a non-linear fashion. As a summary, maps of each frequency band can be used: delta, theta, alpha and beta. If finer frequency resolutions is desired, one has available individual maps of 0.5 Hz steps. This may be necessary if more frequency details are desired, e.g., the alpha peak topography.

The individual maps in 2 Hz steps here give more details, for instance the field at various points of the alpha component, which may show a different spatial distribution than that of the alpha peak. The 4 spectral band maps are a convenient summary, and is a fast way to review the spectral data.

It is important to again discuss the effect of the choice of reference electrode on the FFT. While we have seen that potential maps can be easily transformed from one reference to another, it is not the same in the frequency domain. FFT results are critically dependent upon the reference, and cannot be transformed in a similar manner. In other words, FFT maps based on linked cheek reference are very different from those based on the average reference (Fig. 1 - 51).

Here, at the 7.5-9.5 Hz map, there is no longer an occipital alpha dominance as seen previously. The averaging action has "redistributed" the occipital alpha activity more evenly across the scalp. Similarly, the asymmetry that was present for the theta activity (greater in the left hemisphere) is no longer as clear. Despite these seemingly undesirable effects, there are situations in which average reference is preferred as in some mathematical processing. Furthermore, such maps will always have an evenly balanced colour scale as the zero level will always be in the middle of the voltage range for all points on the scalp at that instant, so that there will be equal amounts of positive and negative values. Whether this characteristic helps or hinders interpretation has to be judged by the individual case.

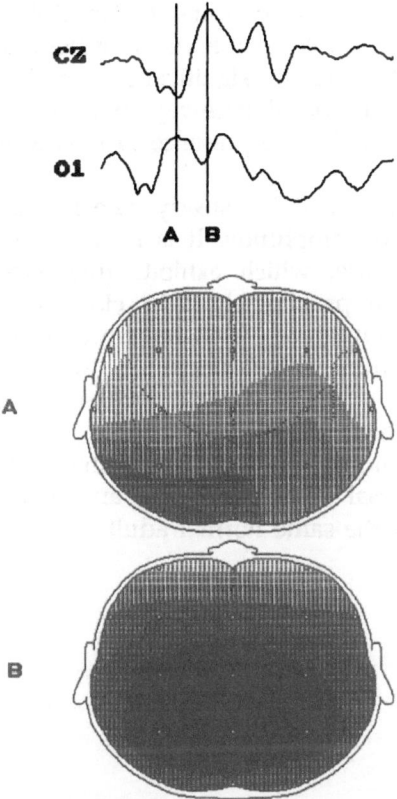

1-55: Flash VEP of same subject. The maps at the peaks of the occipital and vertex response are shown, illustrating occipital and mid-central/parietal peaks. Same scale as Fig. 1 - 54.

The spectral maps can be utilized to show effects of various manipulations; for example, a motor task, mental activity or drowsiness. It can even be used to show a steady-state response to external stimuli such as stroboscopic driving. Fig. 1 - 52 and 1 - 53 illustrate some examples in a normal subject. In the first example (a mu rhythm), the 9.5 Hz component was clearly seen in the bicentral locations, more so on the left. Gentle stroking of the hand resulted in strong attenuation of this activity on the contralateral scalp. The second example demonstrates the effect of visual input, except that this particular example is extremely unusual: visual input from each eye affected the contralateral occipital 10 Hz alpha activity. This child has no symptoms referrable to the visual system, and the observation was an accidental finding. Its significance is unknown, although one may postulate that the visual pathway may be entirely crossed. One contributing factor may be the presence of strabismus in early life.

In contrast to transient EP, steady-state EP can be obtained more quickly, but has lower amplitude. It is most often encountered in the photic driving response, which exhibits frequency sensitivity, being maximally seen at approximately 8-16 Hz, and decreasing beyond that. With a good amplitude, such responses can be clearly seen in the tracing. FFT offers enhancement of this response and allows further quantification. With suitable mathematical preparation, the result can be further subjected to source analysis (Lehmann and Michel 1989). For comparison, Fig. 1 - 54 shows the steady-state VEP due to pattern reversal stimuli at 3 Hz, and a flash VEP in Fig. 1 - 55, both being from the same normal adult.

PART 2

SOURCE MODELLING AND ANALYSIS

2.1 CONCEPTS OF A SOURCE

A powerful framework for topographic analysis is available through the concept of hypothetical sources. This approach relies upon the consideration that active neuronal aggregates can be approximated by discrete sources, each capable of generating a portion or component of the final signal that is manifest at the scalp. This then frees us from the confines of having to define all the neurophysiological parameters pertaining to these neuronal substrate, a task impossible except under the most simple or controlled experimental conditions. We now can concentrate on the characteristics of the signal itself, and make mathematical approximations ("models") of the source configuration.

For example, Fig. 2 - 1 depicts a potential field, which might have been generated by a layer of electrically active cortical neurons. Such a layer is arranged as an ideally orderly arrangement with all neurons aligned side by side, and dendritic trees pointing up (radially). In this simple source configuration (also referred to as a dipole sheet or layer), we can denote the active substrate by a single schematic neuron, symbolised by an arrow pointing in the same direction as the orientation of the dendritic elements. The direction indicates the polarity. The base of the arrow is located at the center of the substrate. How good this approximation is can be measured by how closely the schematic neuron models the geometry of the substrate. In the case where the substrate layer has complex shape, or has other heterogeneous physical and electrical characteristics, such a simple physical model will be poor. These conditions can be present in highly curved gyral areas, or near skull foramina, where the

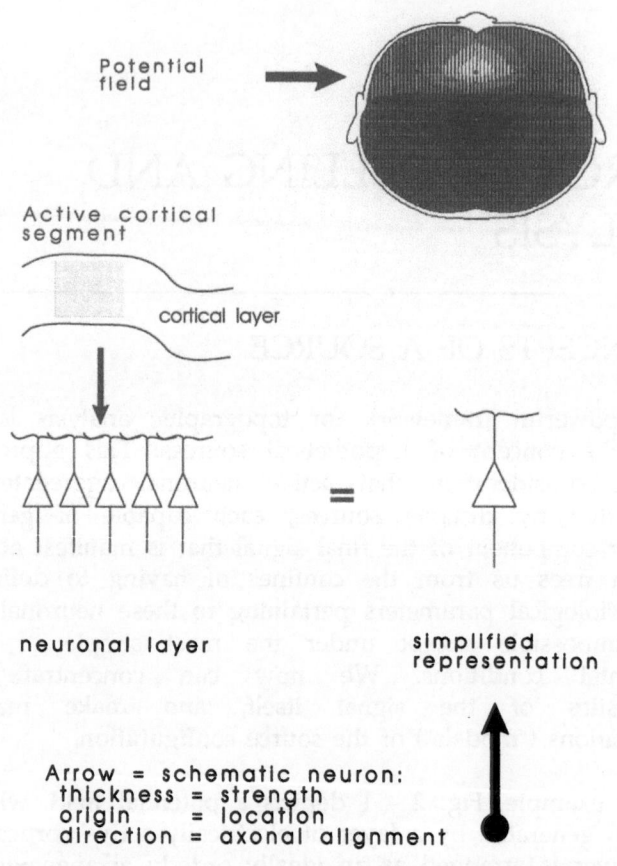

2-1: Scalp potential field representation is depicted on a head diagram at top right. Black = negative, white = positive. The active neuronal elements (generator substrate) are shown as a layer of cells concentrated within a localized cortical segment. The simple geometry of this layer allows simplification to a single "mega-cell", which mathematically can be represented by an arrow.

electrical impedances vary greatly over small anatomic regions. Improved modelling can still be achieved but at the price of having to use a larger number of schematic neurons.

To mathematically describe these physical source models, some assumptions must be made. These are the parameters which constitute the mathematical model: strength of the neuronal sources,

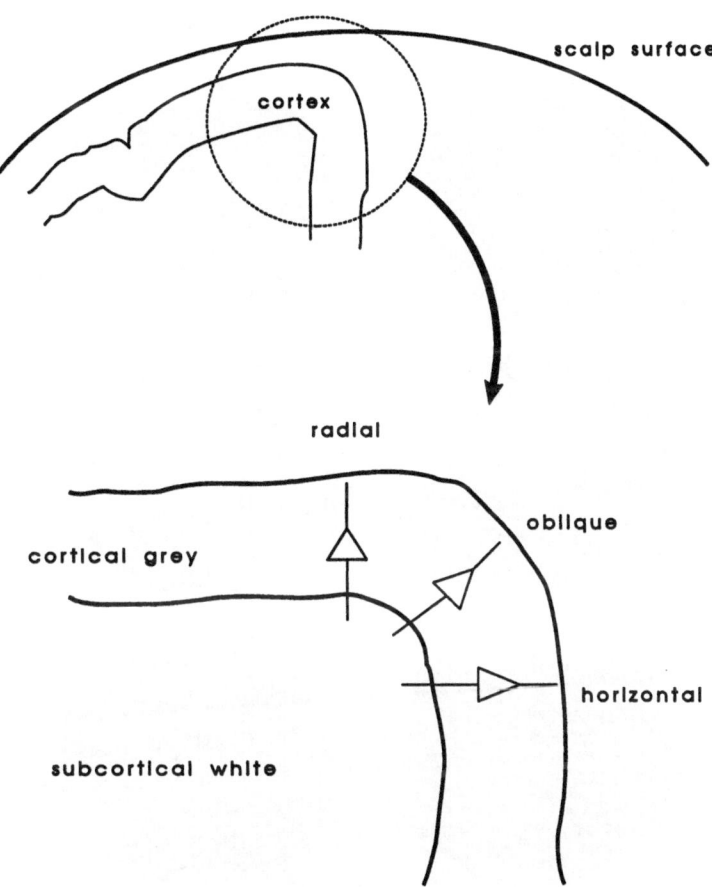

2-2: Source geometry near a fissure. Top: there is sudden orientation change for the elements within the cortical layer near the edge of a fissure. Cells are orientated radially at the flat protion of the cortex, and undergoes a 90 degree change into a horizontal posture within the fissure, which is also tangential to the nearby scalp. This has the effect of rotation.

electrical conductivity of the medium, skull effects etc. A more detailed treatment of this subject will be found in a later Chapter. Let us assume that the electrical effects of these can be represented by a mathematical entity (referred to as "current dipole"), which of necessity must include amplitude and direction information. To simplify discussion, we shall ignore any distortions due to displaying a curved scalp onto a flat surface, bearing in mind that there are

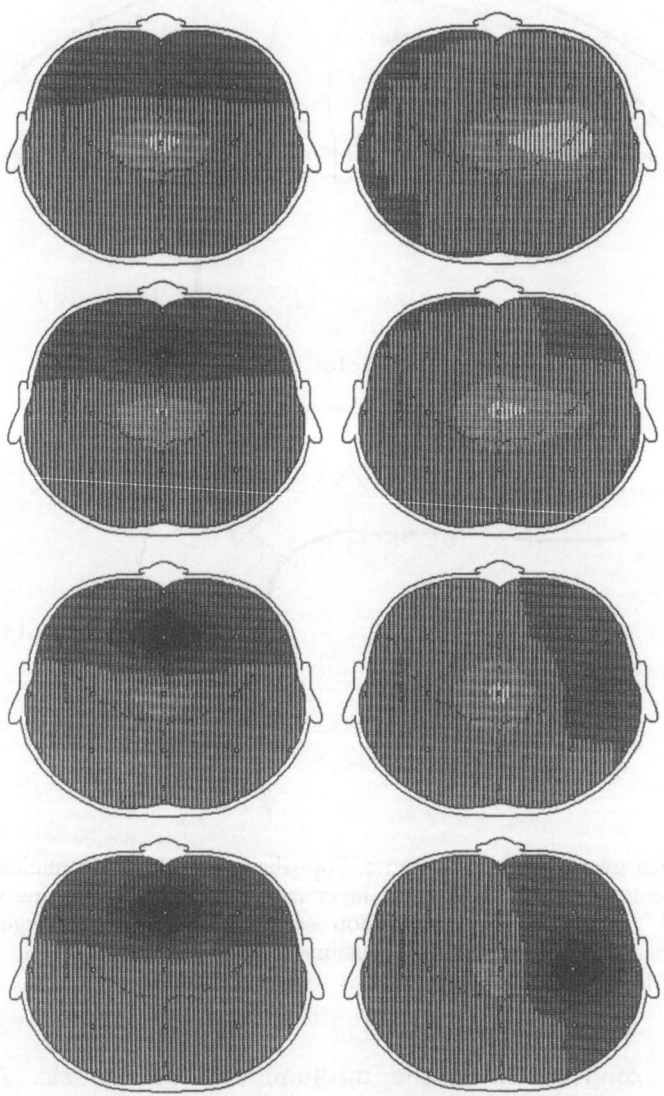

2-3: Source rotation. Fixed source between Cz and Fz, with rotation in the coronal (left column) and saggital planes (right column). The representations of fig. 2-2 are given by the second (tangential), third (oblique) and forth (radial) maps on the left column. White = +, black = -.

methods to correct for these cartographic or projection distortions. We can now explore what happens to the scalp topography with changes in source orientations as might be encountered near a fissure.

Fig. 2 - 2 demonstrates that the orientation greatly influences the recorded scalp topography. The resultant map patterns would be the same if a source of fixed location underwent gradual rotation, with the rotational frequency being low enough for adequate digital sampling. Fig. 2 - 3 shows a series of maps resulting from rotation in the coronal and the sagittal planes. There is obvious similarity in the patterns.

As an added factor, the depth of a source also influences its topography. Firstly, there is the attenuation effect, resulting in decreased amplitude. The effect of volume conduction is more subtle. Fig. 2 - 4 displays the maps from a source located at a point between Cz and C4, at various depths. The orientation is along the coronal plane. As source depth increases, there is a gradual widening of the separation between the positivity and the negativity. A superficial source is associated with a more compact pattern.

The effect of translocation is somewhat easier to understand than that of rotation. This can be seen in Fig. 2 - 5, where each source has the same amplitude, orientation and depth but is placed at different locations. If the source happens to be placed directly underneath an electrode, the resultant maps are identical in shape, with a peak covering a given area centered around the electrode. If the source does not coincide with electrode location, there is the further effect of dispersion due to interpolation, with the resulting map having a distorted peak with larger area (see Fig. 2 - 6 and Fig. 1 - 12). This distortion is due to the limited spatial density of the electrode array.

Such dispersion distortion can be diminished by the use of more electrodes to cover the same area (i.e., higher spatial density), or by the use of more sophisticated (non-linear) interpolation algorithms. The latter is subject to problems discussed in Chapter 1.2. Some promising approaches have been described in the recent literature (Perrin et al. 1987).

It should be added that such anatomic simplifications are often used during routine interpretation of clinical EEG and thus are not a

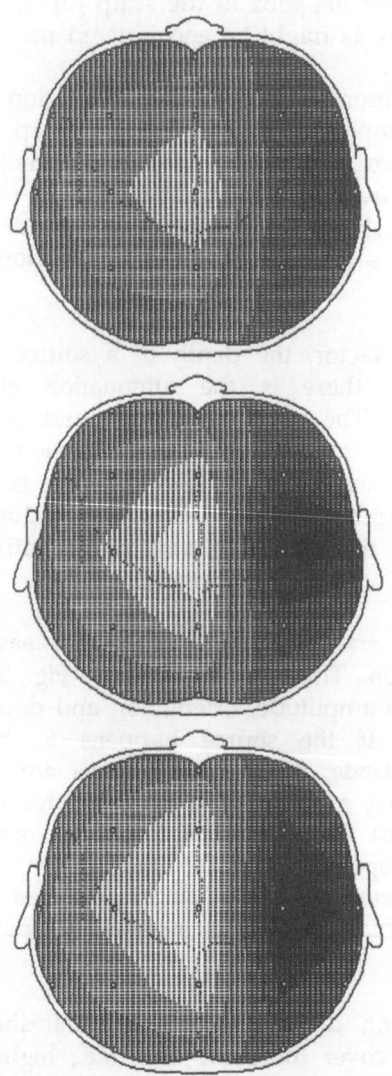

2-4:   Effect of depth. A single source located underneath the midpoint between Cz-C4, but at varying depths. Top: deepest; middle: intermediate depth; bottom: most superficial placement. There is a relationship between source depth and the separation of the positive and negative points on the scalp field.

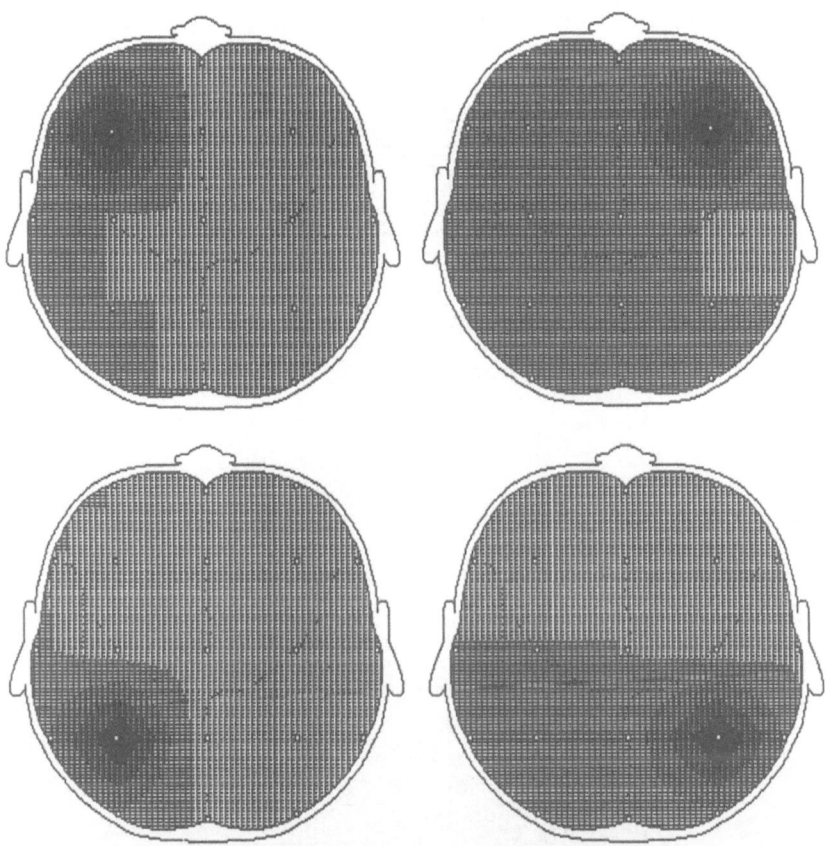

2-5: Source translocation. Identical sources placed directly under the F3, F4, P3 and P4 locations. Note the similarity of the 4 fields, all being circular.

radical departure from daily practice. What is different is the way such simplifications are used. Instead of a relatively vague conceptualization ("feeling") of the generator configuration for a given potential field, we have used an objective mathematical model with well-defined parameters and variables. The results from such a model is based on the recorded potential field at any time point, and can be recalculated at will, for instance using different model characteristics. The human eye of course lacks such fine time resolution, but has the advantage of long experience.

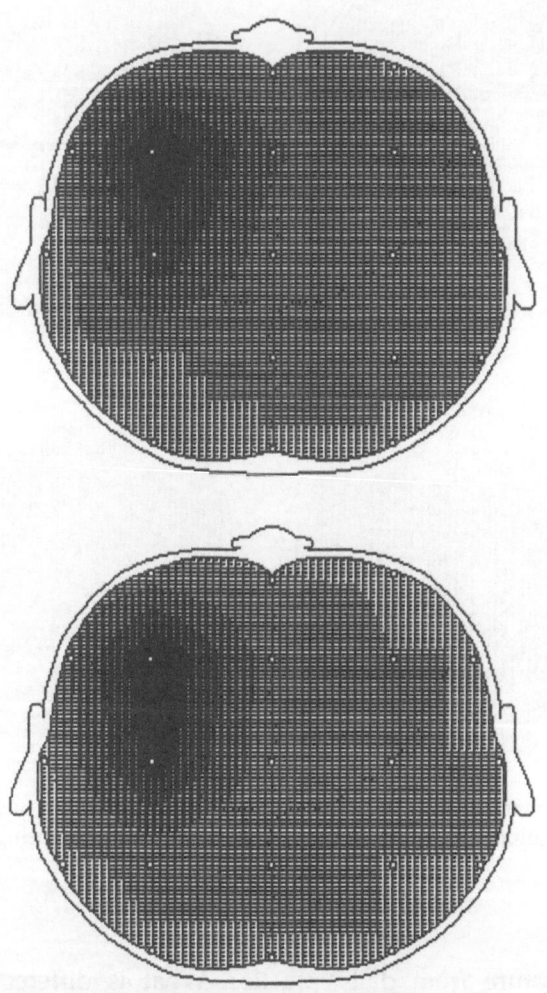

2-6: Dispersion effect due to the source not being directly underneath but slightly posterior to the F3 electrode (top), and even more posterior, at midway between F3 and C3 (bottom). Compare with the top left map of fig. 2-5. The field is not circluar, but distorted and dispersed over a larger scalp area.

To go beyond these initial concepts, another layer of mathematical formalism must be introduced.

2.2 PHYSICAL MODEL

It is necessary to create a mathematically and anatomically acceptable model of the head. First, the presence of the skull constitutes a major impediment to intracranial signals from reaching the scalp. To a lesser degree, other factors contribute: scalp, muscle, hair, meninges, cerebral-spinal fluid (CSF) and of course, brain parenchyma. There are geometric factors to be considered: the shape of the skull, variation in its thickness; presence of skull openings (foramina and sutures); location, size and shape of the ventricles. Finally, there are neural factors: the cortical architectonics; neuronal connection pattern including fasciculi; and presence of pathological excitatory or inhibitory influence.

The majority of these factors are not known to any degree of precision. However, it is still possible to arrive at a physical head model which is in good agreement with observed data.

In a simple model, a spherical cranium is assumed (Fig. 2 - 7). In the first approximation, this single sphere does not provide for various covering layers (skull, scalp, CSF etc), and it assumes a single value for electrical conductivity (a single-sphere model). An improvement would be to allow for separate layers, each with individual conductivities appropriate for bone, muscle/fatty tissue and CSF. Such a three-sphere model is in common use.

Further improvement requires attention to other geometric factors. In order to correct for the presence of skull shape and thickness variations a detailed description is required. Discarding the simple sphere, and using a large number of geometric "finite elements" makes it possible to create a head model which is anatomically accurate to any arbitrary level of precision. Such elements may be simple triangular or square shaped pieces fitted together much like a jig-saw puzzle to form the skull (Fig. 2 - 8). Other sets of elements are used to model the ventricular system and other features. Clearly, the detail achieved can be considerable. The

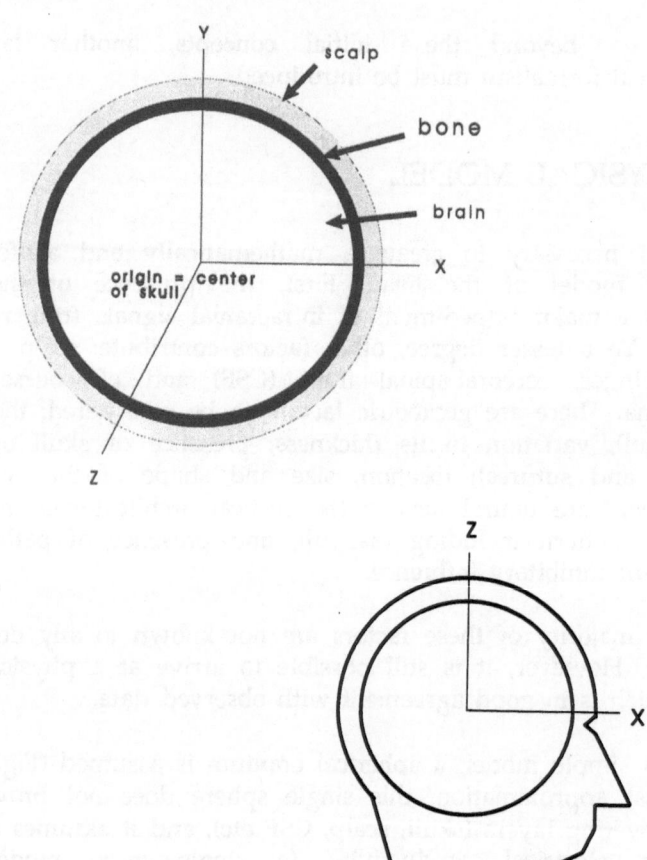

2-7: Physical head model. A 3-sphere model and its position within the head. A cartesian coordinate system is used for location purposes. A realsitic skull shape has not been taken into consideration.

price to be paid is the sky-rocketing mathematical complexity, and the attendant computational requirements.

The stage is now set for seeing how a source would fit into a head model. Fig 2 - 9 shows a cartesian coordinate system used for localization purposes. The origin is usually at the center of the sphere, which anatomically corresponds approximately to the midpoint of the inter-auricular line. The orientation of the source can also be described by its angle relative to a chosen axis (e.g., the X axis in Fig. 2 - 9), and its elevation from the X-Y plane. Finally, its

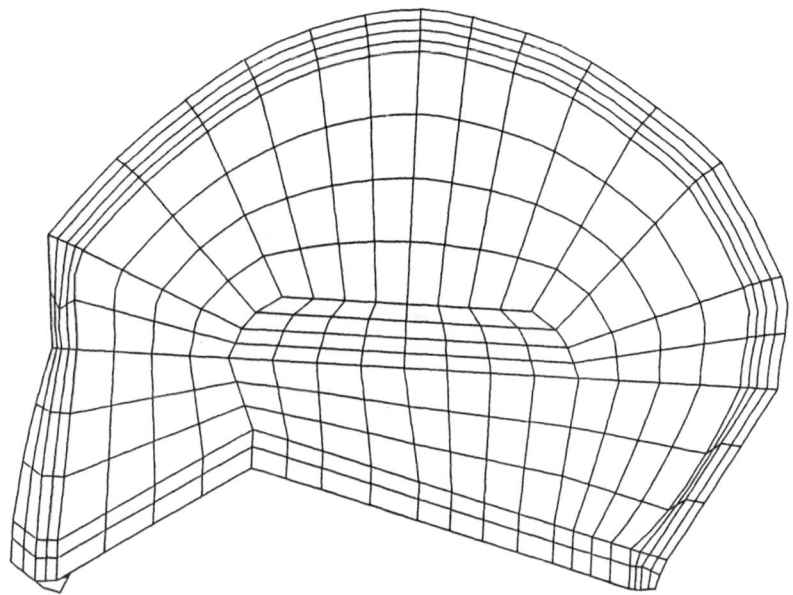

2-8:   Finite element head model composed of a large number of simple 4-sided elements of different sizes and shapes, stretched to conform to the skull shape. It is anatomically more accurate than the previous spherical model (figure courtesy of Dr. Paul Nunez).

amplitude is denoted by the length of the arrow. Thus there are 6 unknowns for each source: 3 for location, 2 for angular relations, and 1 for amplitude. If no constraints are imposed, there are then 6 independent degrees of freedom (d.f.).

As an example of constraint, if the location of the neuronal generator is known, the source location can be fixed. This reduces the d.f. to 3. The advantage is that better mathematical results can be obtained. This will be elaborated upon.

If there are 2 neuronal generators, 2 sources are required, giving 12 d.f. However, if assumptions of hemispheric symmetry can be made for their locations (e.g., symmetrical sources in homologous cortical regions), the 2 sources can be described by fewer location variables. This further decreases the d.f. to 9: 3 for location of source A, and 2 each for orientation, and 1 each for amplitude.   The location

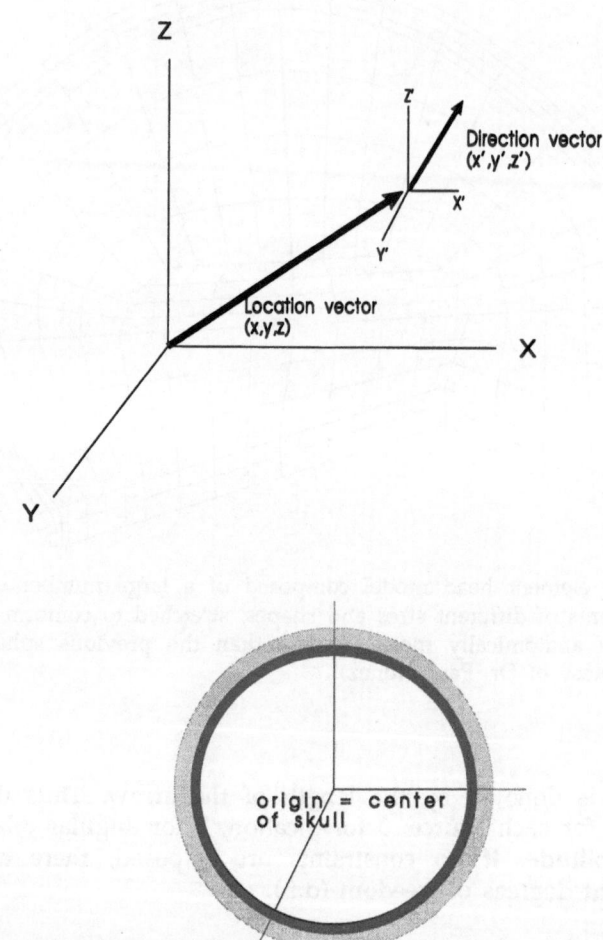

2-9: Coordinate system. A source location is defined by a first vector (x,y,z), referenced to the center of the spherical skull. Its direction is defined by a second vector, using a second set of axes with origin at the point (x,y,z). The length of the location vector is the radius of the source from the center of the sphere, while the length of the direction vector is the magnitude of the source.

for source B is purely dependent upon that of A and can thus be omitted. This is an example of location constraint (Fig. 2 - 10 middle).

A more useful example is to disallow extra-cranial locations. It is also possible to arrange for 2 sources to have parallel orientations (Fig. 2 - 10 bottom). An important objective in all these examples of constraint is to allow a restricted mathematical solution. This will be discussed in the next chapter.

Let us see how such a source model works in practice. If a source of known characteristics (location, orientation and amplitude) is assumed, and the physical head model is appropriate (correct values of head geometry and electrical conductivity), we can calculate the scalp electric field precisely. The only errors are from the assumptions made in the modelling process. These errors, however, are constant. Therefore the prediction of scalp potential field is straightforward. This process is aptly named the *forward solution*.

A general consideration of how some of these model assumptions affect the forward solution is now discussed.

A single-sphere model will estimate a source to be more superficial (less centric or having a larger "radius" from the origin) than a 3-sphere model, since the attenuating effect of the skull (being the main contributor) has not been fully accounted for. As an approximation, the effect of skull attenuation can be accounted for by simply decreasing the estimated radius (Schneider 1974) by a factor of approximately 0.6. To be more precise, this correction factor should not be a constant, but should vary depending on skull thickness. It can be appreciated that the skull model would benefit greatly from detailed geometric information as can be provided by an MRI head scan.

If the source is not discrete, but instead is extended like a dipole sheet, several sources must be used to mathematically describe it correctly (Fig. 2 - 11). This is even more important if the dipole sheet is not flat. The situation deteriorates drastically if the dipole sheet is closed around itself as in a cylinder or sphere. In this circumstance a closed neural network results, and it is well-known that signal cancellation occurs for distant recording electrodes.

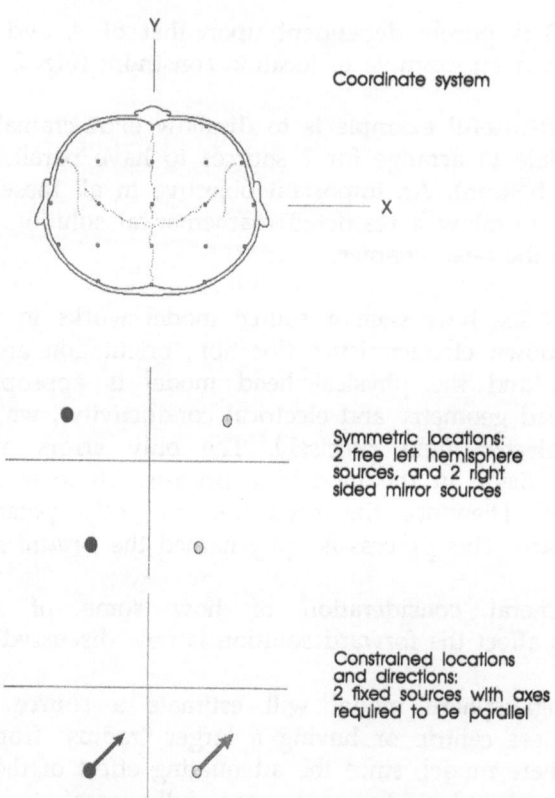

Y

Coordinate system

X

Symmetric locations:
2 free left hemisphere
sources, and 2 right
sided mirror sources

Constrained locations
and directions:
2 fixed sources with axes
required to be parallel

2-10:   Constraints. Top: coordinate system. Middle: the 2 left hemisphere sources are free, while the 2 right hemisphere sources have exactly mirroring locations. Bottom: the right source is dependant upon the left for both location (mirror image) and direction (parallel).

Multiple sources may also be simplified under certain favourable condition. If 2 adjacent sources are orientated in parallel, a single equivalent source located between them, with amplitude being the algebraic sum, provides adequately representation (Fig. 2 - 12). There is an implicit assumption that these 2 discrete sources are electrically tightly coupled, thus behaving synchronously. Should this synchrony not be present, the single equivalent source representation would be erroneous. The chronologic behaviour of sources provide important clues to how many independent sources exist in a given dataset.

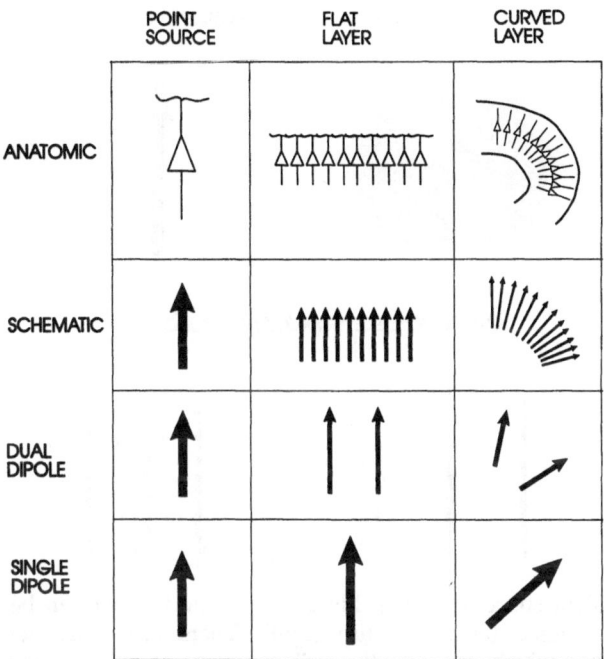

2-11: Presentation of different source types: hypothetical point source (left column); flat layer (middle column); curved layer (right column). For each source type in the top row, its schematic representation is seen in the next row. The bottom 2 rows are the simplified representations using single and dual dipole models.

The foregoing discussion applies to sources whose locations are constant over the time interval of interest. To incorporate movement, location parameters are incrementally changed from (x_0, y_0, z_0) to (x_1, y_1, z_1), and the forward solution re-calculated. The orientation and amplitude can be similarly changed.

The approach taken in this chapter is merely illustrative. Many practical considerations and details have of necessity been omitted. To include sufficient material for a realistic head model would be outside the scope of such an introductory volume. The interested reader is referred to the many excellent sources in the literature for more information (Scherg 1985a, 1985b; Meijs et al. 1987; Fender 1987; Nunez 1989).

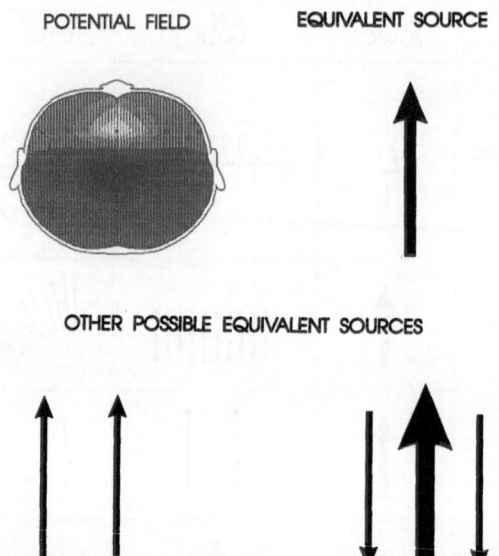

POTENTIAL FIELD   EQUIVALENT SOURCE

OTHER POSSIBLE EQUIVALENT SOURCES

2-12: Equivalent sources. A map illustrated at the top left can be represented by a single source (arrow at top right). There are other source models (configurations) which can equally well represent the same map and which cannot be distinguished from the single source model on mathematical grounds.

2.3   INVERSE SOLUTION

In the previous chapter, we had assumed total knowledge of the source characteristics. Together with an appropriate head model, the forward solution predicted what we would observe (measure) at the scalp. In practice, what we have is the scalp electric field, from which we wish to derive the unknown source characteristics. Under certain conditions, the unknown sources can be estimated by the process of *inverse solution*. This is a complex process, with no guarantee that the derived solution is physiologically correct. Worse, there are infinite possible solutions which are mathematically correct: i.e., a "non-unique" nature.

The basic steps can be stated as follows:

1.   Assume a physical model with a fixed number of sources

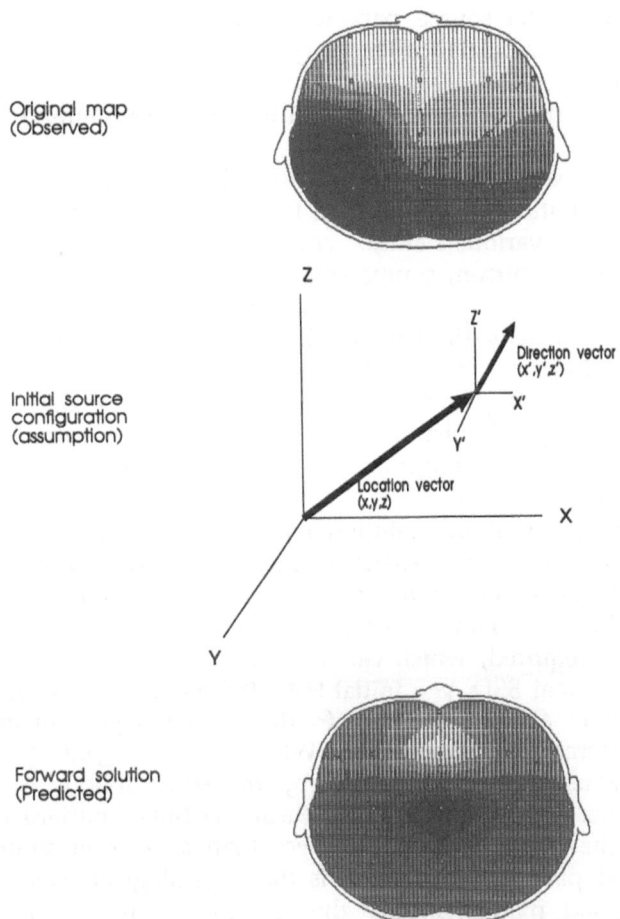

Original map
(Observed)

Initial source
configuration
(assumption)

Forward solution
(Predicted)

2-13: Source estimation. An observation at one instant is shown as the top map. To start the DLM procedure, initial source location and direction vectors are chosen, usually arbitrarily. The field generated by this first guess is calculated (forward solution).

2. Input the scalp electric field "O" (= observed) at a given instant of time
3. Assume initial source parameters (location, orientation and amplitude)
4. Calculate the forward solution "P" (= predicted)
5. Calculate the difference "D" between "O" and "P" across all electrode positions

6.   Each of the source parameters is adjusted systematically, then "D" is recalculated after every adjustment by repeating steps (4) and (5)

7.   Trial-and-error iterative search to find the minimum "D"

By repeating this procedure for all time points, a complete set of source solution is found (see Fig. 2 - 13 and 2 - 14).   Such a procedure has various names: source localization, dipole localization method (DLM), current source solution, etc.

"D" is usually the sum of the voltage difference squared ("sum squared difference" or SSD) at each electrode. The minimum is deemed to have been reached when consecutive "D" decreases by less than a pre-assigned tolerance. The values of the 6 source parameters corresponding to this (optimal) "P" is the solution.

Having found the solution which has the least difference from the original map by an exhaustive search (Fig. 2 - 15), the question is now what the "goodness of fit" (g.o.f.) is. This encompasses at least 2 aspects. The first and easier is the mathematical sense. A definite criterion is required, which can be related to the amount of residual SSD. If x = final SSD, y = initial SSD, the g.o.f. = $(1 - x/y) \times 100\%$. A good fit gives a low percentage (< 10%), while a poor fit gives a high residual percentage (>> 10%). While such a g.o.f. parameter is commonly used, it is influenced by the serendipitous choice of the initial guess. A good initial guess (near the final solution) will yield a value of the initial SSD much lower than a poor or unlucky guess. The second part of the question is the physiological sense. Regardless of how good mathematically the solution is, no statement can be

2-14: Inverse solution by iterative search. The calculated map (top map) from the initial source solution (top arrow) is compared with the recorded map (Fig. 2-13 top) and the difference (error) noted. A small change (perturbation) is applied to the source and the calculations repeated. If this error is greater than the previous, the most recent change is discarded and a fresh change applied. The source configuration associated with the smallest error is always retained.

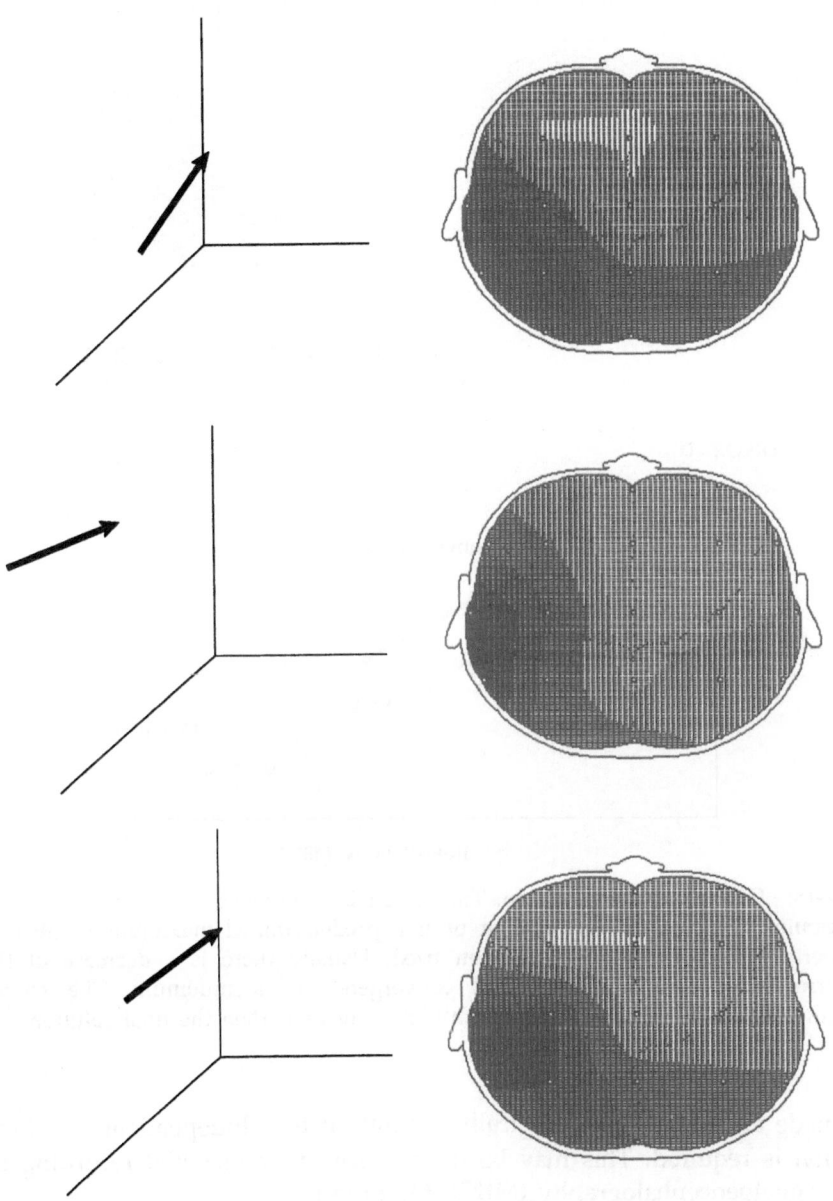

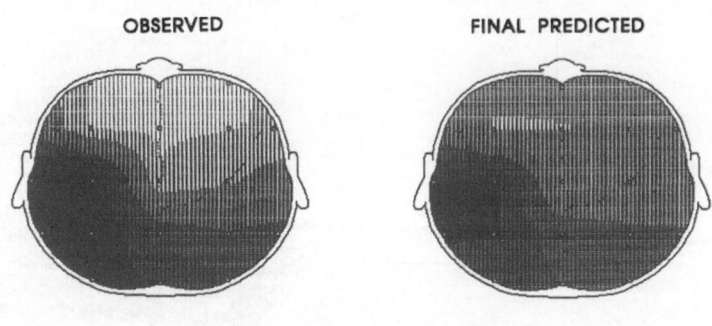

OBSERVED   FINAL PREDICTED

MAP REPRESENTATION BY THE SET OF 6
SOURCE INDICES = (X,Y,Z, X',Y'Z')

ERROR 'D'

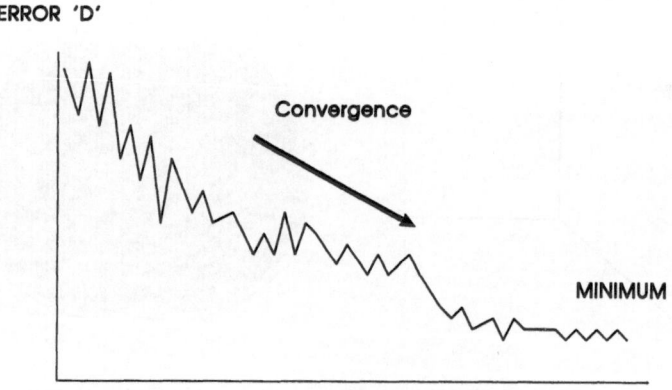

Convergence

MINIMUM

ITERATION NUMBER

2-15: Convergence to solution. The search is terminated when the error term cannot be decreased any further or if a predetermined maximum number of perturbations (iterations) has been tried. Usually there is a decrease of the error with each iteration, until convergence to a minimum. The source configuration associated with the minimum error is then the final solution.

made as to how physiologically accurate it is.   Independent verification is required. This may be in the form of intracranial recording or magnetoencephalography (MEG), for instance.

A common problem is that of extracranial solutions. This can be corrected by the use of constraints. This is achieved by accepting only intracranial (i.e., physiological) solutions during the search. This may be generalized to allowing solutions from one hemisphere, or some other defined brain region. The rationale for this may be some a priori knowledge of source location. Indiscriminate use of constraints will of course yield erroneous solutions.

Another problem which may be encountered is related to poor convergence. During each iteration, a minima of the SSD is sought within the search area, being a fixed and limited distance from the current location. As there are a total of 6 parameters (6 d.f.), this constitutes a search in 6 dimensional space. It is possible that the algorithm converge onto a local minima, which is not the global minima. Fig. 2 - 16 illustrates this situation in 2 dimensional space.

The problem of non-uniqueness of the inverse solution is more serious. For a given potential map, there are many solutions which mathematically fulfil goodness of fit criteria, but do not resemble each other. The judicious use of constraints narrow the field somewhat. Fig. 2 - 12 illustrated examples of possible solutions to a given observed potential field.

This non-uniqueness situation is worse with multiple source models, which assume 2 or more sources in the configuration. A 2-dipole model is much more difficult to handle properly than a single-dipole model. For a given set of data, a 2-dipole solution is likely to be mathematically superior than a single-dipole solution, because the additional degrees of freedom allow a better mathematical fit. Despite the mathematical superiority, there may be even less assurance that the 2-dipole solution is physiologically correct.

A simple mathematical procedure to estimate the nearest scalp location overlying a single source (the "focus", Fxy) is illustrated in Fig. 2 - 17. It is based on the localization of the source to a point between the positive and negative peaks of the potential field. It is also closer to the peak with the higher amplitude. This approximates the position of the greatest voltage gradient. Even if there is no dipolar field pattern present this method still works, as the calculated Fxy location is then the same as that of the single polarity peak. Fig. 2 - 18 explains why this is so, by displaying how an intracranial source can be projected onto the scalp, the scalp position of which can be described by the 2 dimensional coordinates (x,y).

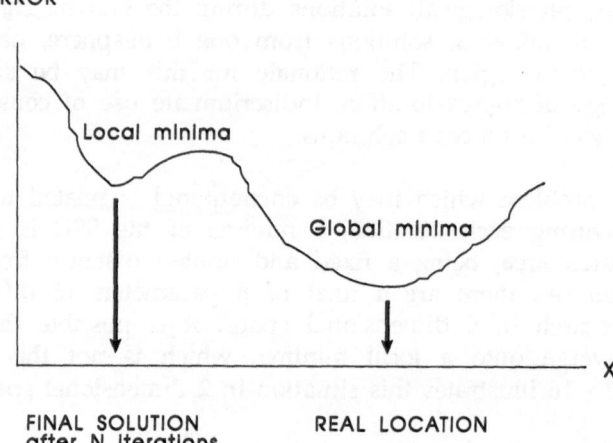

2-16: Local vs global minima. Convergence may be complicated by solutions which shows a local minima of the error after a short search, thus yielding a seemingly good solution. There may well be another better solution elsewhere (the global minima) which was not obvious, and which might have been found after a longer and more exhaustive search.

2.4 STABILITY OF DIPOLE SOLUTIONS

If DLM was applied to an EEG segment of say 256 time points containing a spike, the resultant 256 individual source solutions can be superimposed and displayed as in Fig. 2 - 19. Clearly such indiscriminate data analysis is confusing. It would be much more instructive to selectively display the solutions, perhaps sequentially. The objective would be to discern a pattern of stable solutions, where the location, direction and strength do not change much within a short time window. It may then be reasonable to assume that during this stable time period the underlying generator(s) was adequately described by the model, and not much variation (spatial or temporal) of the generator configuration took place.

A formal measurement of the fluctuation of location and direction parameters can be arrived at as follows. Let us assume that the location is expressed in cartesian coordinates, with the origin at the center of the spherical head. The direction is expressed similarly as a cartesian vector with origin at the dipole location, as before.

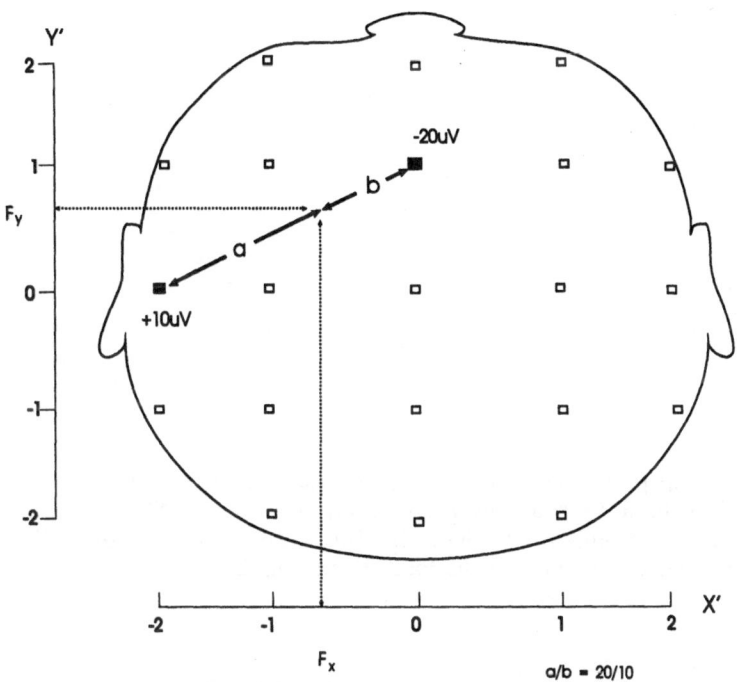

2-17: Source approximation using the "estimated focus" Fxy. The scalp is projected onto a flat plane, the location (or "focus") nearest to a single source may be simply estimated to be on the line joining the two opposite polarity peaks. If one treats the 2 voltages as weights, it is in fact the baricenter (or center of gravity). In the event that there is only 1 peak, then the focus correspond to the location of that peak.

There are thus 6 dimensions, representing each of these 6 coordinates. The fluctuation of each coordinate over successive time points can be described by its variance over a short time window (e.g., 5 time points). An aggregate estimate of stability in both location and direction can thus be calculated (stability index, SI):

$$SI = \log [1/(VAR_L * VAR_D)]$$

where VAR_L, VAR_D are the variances over the 5-point grid for location and direction respectively (see Fig. 2 - 20). Of course, these 2 components can be used separately. The logarithm ensures a range of

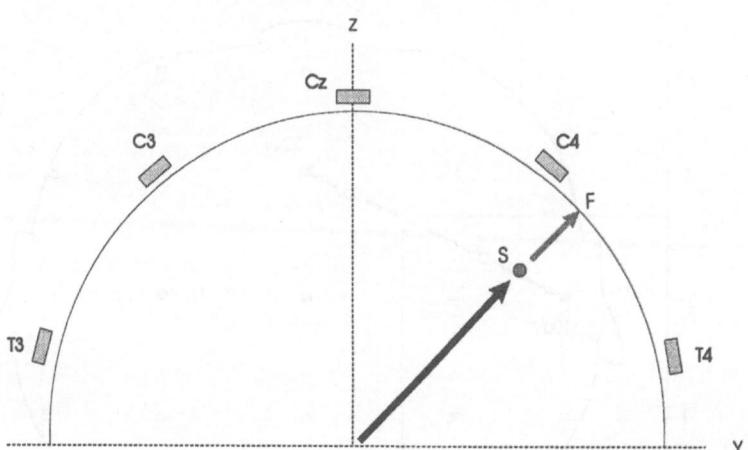

2-18: Scalp projection of a source (the "focus"). A coronal section of the head model is used to show how an intracranial source "S" (in 3 dimensional coordinates) is projected to the scalp. This scalp location "F" (in 2 dimensional coordinates) corresponds to the focus Fxy of fig. 2-17.

usual SI values from 0 to 7. A large value corresponds to very stable source solutions, whereas small or negative values correspond to poor and fluctuating mathematical solutions.

In Fig 2 - 21 and 2 - 22, a simple criterion based on the SI value was imposed, and only those solutions which have an SI exceeding a threshold value were displayed. By this simple use of stability analysis, we have objectively selected stable source "windows" for display, while rejecting those unstable solutions. This of course provides another degree of data reduction. The same results can be obtained visually, by patiently reviewing all DLM results, noting the time periods where the solutions were stable. The same stable "windows" would result. Such an orderly display of DLM results facilitates the identification of locations of stable solution sets, and also the formulation of hypotheses pertaining to the dataset under exploration. This method has been described in detail elsewhere (Wong 1989).

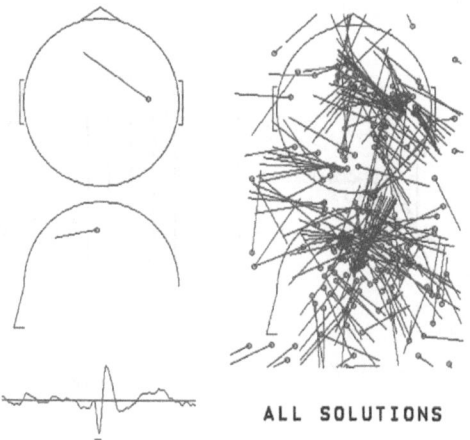

ALL SOLUTIONS

2-19: DLM display. Left: source solution at the negative peak (apex) of a spike, showing the vertex view (top), left side view (middle), and the 1.28s spike segment tracing at one channel (apex marked by short bar); right: superimposition of source solutions at all time points of the spike segment.

2.5 DATA CHARACTERIZATION

This chapter will synthesize several methodologies covered thus far so as to provide a rationale for their use.

One of the difficulties associated with topographic analysis has to do with how one handles the derived data, particularly maps. It is not immediately obvious how maps can be compared with each other, or statistically described. Of fundamental importance is the concept of map characterization. Ideally the thousands of picture elements ("pixels") of a map can be appropriately reduced to one or a few parameters. These "map characteristics" may be subject to parametric or non-parametric statistical analysis, deriving group descriptors or discriminators for between-group classification. These characteristics should be few in number, easy to understand, and can be expanded to yield the original dataset (i.e., map) with minimal errors. To be easy to understand, they should relate to the electrophysiology; to be re-expandable, and they should adhere to a mathematical model.

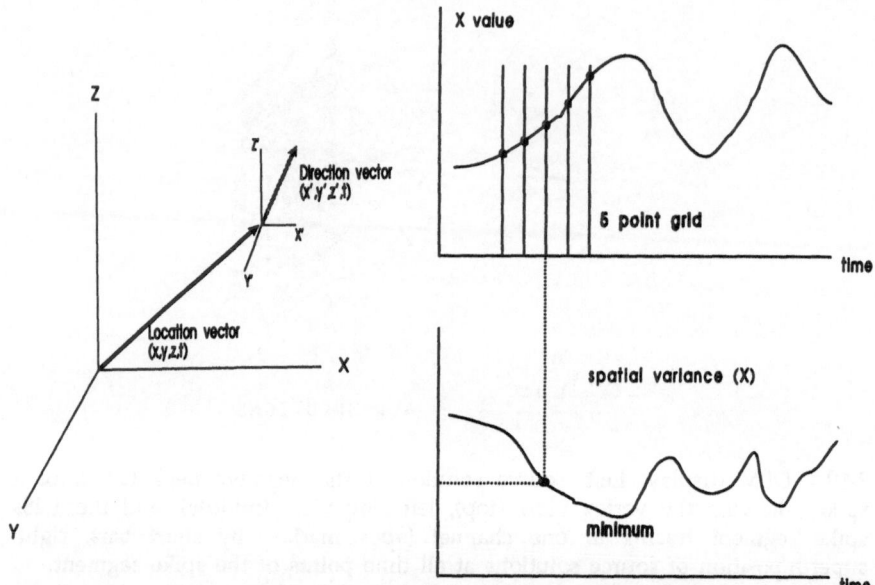

2-20:   Stability Index. Left: a source described by the 2 conventional vectors; right top: tracing of the x-coordinate vs time; right bottom: spatial variance over time, showing a minimum as calculated by the 5 point grid.

Fig. 2 - 23 presents an isopotential map and several possible descriptors. The simple peak display may be perfectly adequate under some circumstances (Lehmann et al. 1988), whereas other circumstances may require the accuracy of DLM solutions (Wong and Weinberg 1988). Any of these methods is sufficient if the objective is to describe which hemisphere a spike originates from. If we need to discern the subtle differences between the spikes from 2 clinical groups, the more precise descriptors are required. Given an unknown spike sample, we may wish to predict which clinical group it belongs to with low error rates. Such a classification procedure requires the identification of characteristics which have high discriminating power. In this vein, the issue of anatomic accuracy of DLM solutions is not necessarily pertinent. The source parameters are merely used as classification indices or descriptors of the data. As an illustration to how one might develop descriptors for a given problem, the following example is used.

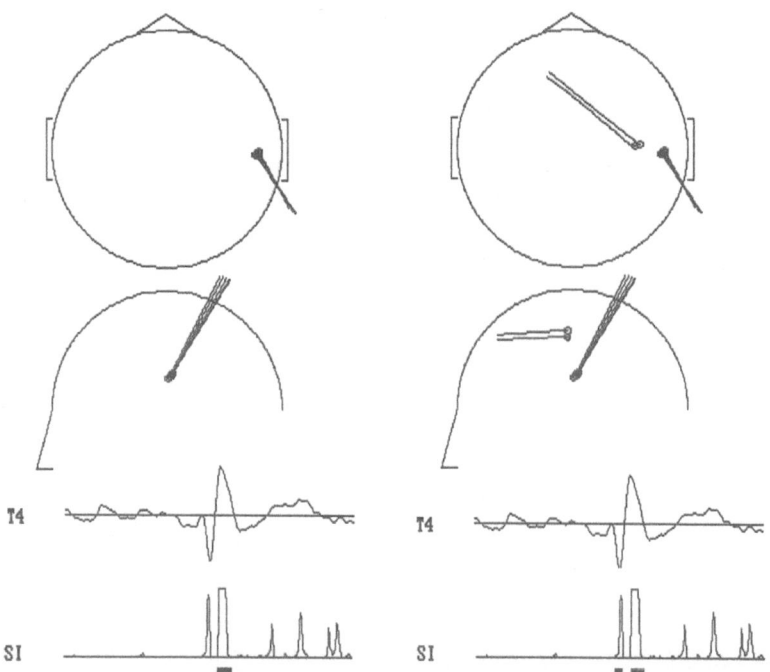

2-21:   Source fluctuation of rolandic spike. Left: the most stable source solution (pointing to the right posterior and up) is near the positive peak of the EEG spike (bar on the SI tracing). Right: slightly less stable (lower SI) solution at the earlier negative peak (shorter bar), pointing more anteriorly and to the left.

Let us consider spikes segments obtained from patients with benign rolandic epilepsy of childhood (BREC). Clinically, there are children who have typical Sylvian seizures (Lombroso 1967) but who also have neurological findings, thus not falling within the traditional definition of BREC. They may be called atypical BREC (Wong et al. 1988). The objective is to seek descriptors of the EEG spike data which can discriminate between the typical and atypical groups. The data consists of spike segments collected from these 2 groups. Fig. 2 - 24 gives some representative samples of the EEG fields at the spike apex. Visual analysis of the entire dataset did not yield obvious patterns of differences between the 2 groups. We then have to sys-

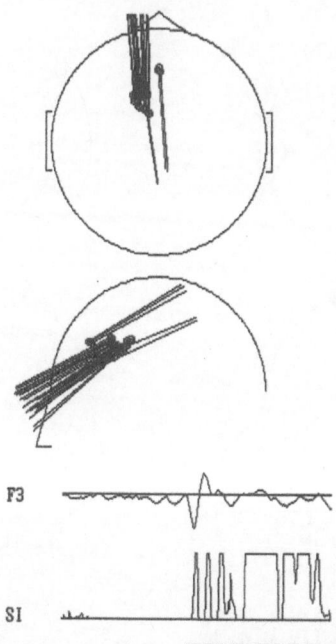

F3

SI

2-22:   Stable solutions. Same format as in fig. 2-21: display of those stable source solutions exceeding a certain SI threshold, seen to occur over the period marked by the bar.

tematically select descriptors and test their classification ability (an "explorative study").

Both chronologic and topographic descriptors will be included in this exercise. As the objective is to demonstrate the methodology, rather than to prove something scientifically, some details will be omitted. The interested reader is directed to the original publication (Wong et al. 1989). Due to a limited amount of data, the entire dataset will be subjected to exploration. In this sense, it is known beforehand which clinical group each spike is associated with. The object is to find the parameters which can best differentiate the 2 groups. In an exhaustive manner, a first parameter is selected by software which best segregates or partitions the data into the 2 clinical groups as correctly as possible. From among the remaining parameters, the next best is selected, again with the criterion that the

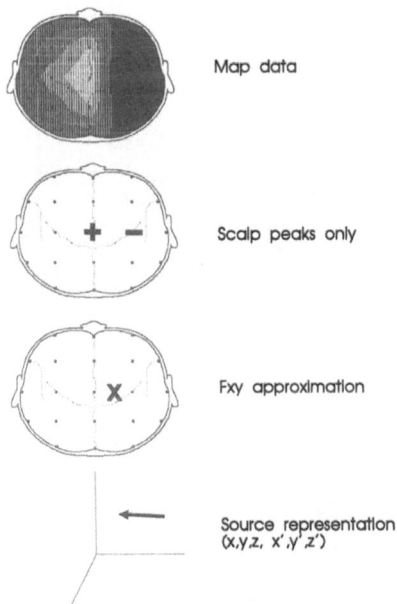

Map data

Scalp peaks only

Fxy approximation

Source representation
(x,y,z, x',y',z')

2-23:   Map characterization. The map at the top can be represented by its
scalp polarity peaks, the calculated focus, or the calculated source solution
(respectively shown below). All are convenient descriptors of the original
map, but with varying degrees of sophistication and fidelity.

resultant groups be as homogeneous as possible. This process is
stopped when there is no further data, or when the aggregate correct
classification rate is high enough. There are formal statistical
parameters which help determine this end-point, or alternatively,
indicate if there is no good classification possible. The mathematical
basis of this technique is described in Chapter 4.5.

Of the original 10 parameters, the chronologically-based ones
were not found to be useful. By using only the amplitude and
location of the spike negativity, and the abscissa value of the
estimated focus, Fx, an overall classification rate of approximately
80% was achieved. The result is a simple "classification rule" which is
physiologically based and easily interpreted. Restated, these 3 derived
parameters were reasonably good in characterizing the data for our
purposes.

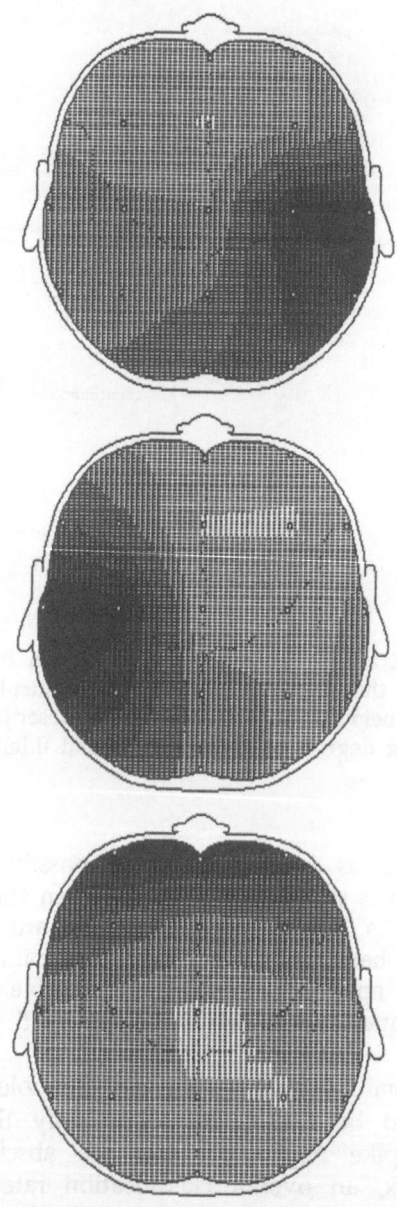

2-24: Example of BREC spike maps recorded from 3 different patients. Each shows the spike field at the negative spike peak (apex). The first two are obviously temporal-central, while the third is not so clearly described, being midline and having an anterior-posterior gradient.

This is an example of pattern recognition. By noting the location and amplitude of the negative peak, and how lateral the estimated "focus" was, a pattern (or template) associated with the typical BREC spike is established, with which we can rely upon for predictive purposes. Such a classification rule is of course only as accurate as the dataset used in its definition.

In another exploration, the dataset were subjected to DLM analysis, using a single-dipole model. The 6 source parameters together with the source eccentricity (i.e., radius, being the square root of the sum of the squares of the 3 location coordinates) were similarly subjected to exploration. It was found that the radius provided the best classification capability. Physiologically, it may be interpreted that the depth of the generator was an important discriminator between BREC and atypical BREC. In this case the secondary derived parameter, radius, was the most efficient characteristic.

PART 3

MAGNETOENCEPHALOGRAPHY

3.1 INSTRUMENTATION

Although theoretical physics predicts that currents in biological media result in magnetic fields, the predicted fields are so small that until only recently they were unmeasurable.

One of the first biomagnetic measurements was of fields associated with cardiac function, measured by Cohen et al. in 1970. Coil magnetometers of the kind they used are usually not sensitive enough for the detection of brain function which is an order of 10^{-4} smaller than fields produced by the heart. Consequently it was not until a Josephson Junction was incorporated into a Superconductive Quantum Interference Device (SQUID) that magnetometers with the required high sensitivity were available for the measurement of brain magnetic field. The SQUID, used as an ultrasensitive magnetic flux detector, has a superconducting ring with a weak link with a cross-sectional area of only a few microns. When an electric current is induced in the SQUID the current normally flows without resistance (i.e., the material is superconducting). However the current density is much greater where the current is "pushed through" the weak link. When the current density becomes significantly greater the superconducting SQUID becomes temporarily normal conducting (the Josephson effect). The rate at which the SQUID changes from its normal conducting state to its superconducting state is proportional to the external magnetic field. This mechanism provides a very sensitive measure of changes in magnetic flux induced in the flux transformers.

To achieve superconductivity, the sensor has to be cooled to the temperature of liquid helium (4.2 degrees Kelvin). In order to accomplish these low temperatures the SQUID is immersed in liquid helium inside a special container (dewar). The rf-SQUID is a superconducting ring with one Josephson junction (weak link) in it. The dc-SQUID has two weak links in the ring. The sensing coil is part of a system that transfers magnetic field flux to the SQUID, and is closest to the brain. Very small currents are induced in the coils and those currents are transported to the SQUID. In a magnetometer there is only one coil. In gradiometers several coils are used, each of which is separated from the other by a distance (the baseline). Gradiometers are one stage of the process of extracting signal from noise, that is the process of measuring small signals resulting from brain activity in the midst of much larger signals from the surrounding environment. The principle is very much like that of a differential amplifier, used in EEG devices. The differential amplifier is one that measures the differences between signals. The generator presents a different signal to each side of the amplifier whereas a very large signal is relatively less different.

For magnetic signals the strength falls off as the cube of the distance. Therefore a small signal will be much more different at two coils separated by some distance (the baseline) than a large signal. A first order gradiometer has two coils that are wound in opposite directions. The currents produced in one coil are subtracted from the currents in the second. Since the currents in the two coils are more similar for a large field than for a small field, the difference between the two coils is *smaller* for the large field than for the small field. In this way a small signal can be distinguished from the large fields that result from surrounding environment which are not of interest and hence constitute noise (Fig. 3 - 1).

The first order gradiometer has 2 sensing coils (Fig. 3 - 2), a second order gradiometer has 3 sensing coils and is more insensitive to homogeneous external magnetic fields, i.e., noise. A third order system has the best capability of discriminating signal from noise. In general, the higher the order of the gradiometer the greater is its capability of distinguishing very close signals from distant signals.

In order to measure the distribution of magnetic fields over the head with instruments that do not cover the whole head, measurements have to be repeated several times. The sequential measurement at many locations is not only time consuming but constrains experimental paradigms to those which assume that the

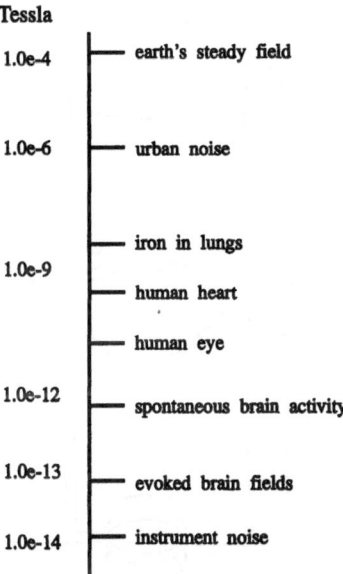

Tessla

- 1.0e-4 ── earth's steady field
- 1.0e-6 ── urban noise
- 1.0e-9 ── iron in lungs
- ── human heart
- ── human eye
- 1.0e-12 ── spontaneous brain activity
- 1.0e-13 ── evoked brain fields
- 1.0e-14 ── instrument noise

3-1:   Relative strengths of magnetic fields.   Note that the level of evoked fields measured from the brain is close to the intrinsic level of instrument noise.

brain is processing the same inputs in the same way during successive measurements (i.e., steady-state). Multichannel instruments, presently up to 37 channels have been recently developed and there is now under construction a 60 channel instrument (Fig. 3 - 3).

Because SQUID-magnetometers must be immersed in liquid helium the physical configuration and size of the measurement instruments make the typical system bulky and unwieldy. Typically, the diameter of the dewar tail is between 5 and 10 cm and the height of the dewar at least 50 cm. The liquid helium container weighs at least one kilogram and cannot be freely tilted. A typical distance between the main sensing coil and the skin is of the order of one centimeter. The instrument is sensitive to vibrations and to movements and it has to be well supported. Depending on the size, the dewars stay cold between one day and about one week.

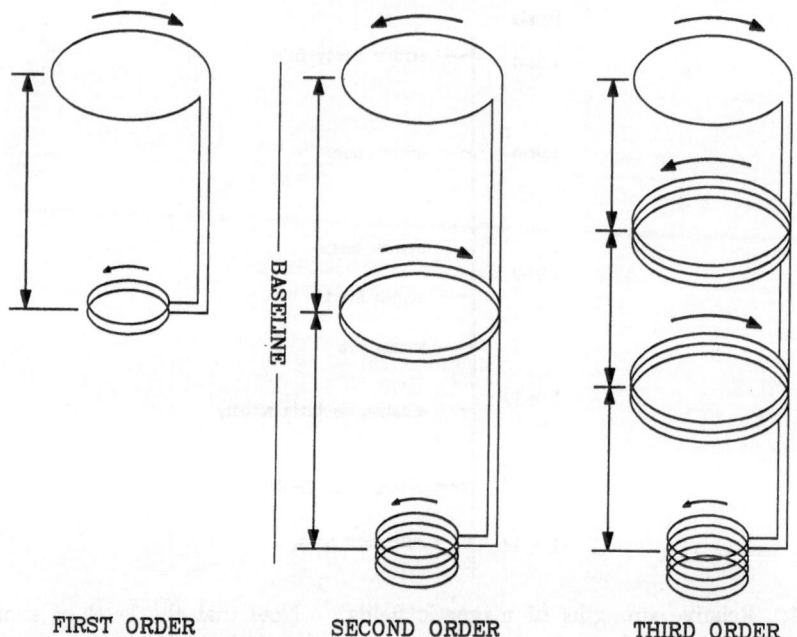

FIRST ORDER   SECOND ORDER   THIRD ORDER

3-2: Gradiometer configurations. A third order gradiometer has the greatest capability of extracting small signals in the near field from large common mode signals in the far field.

When recording magnetic fields which are as small as one billion times smaller than the earth's magnetic field artifacts are a significant problem. Third order gradiometers have the greatest rejection of noise generated from far fields. However, most multichannel systems utilise 2nd order gradiometers and consequently require a shielded room. The shielded room is usually constructed of aluminum and numetal and must be mounted in order to reduce its vibration (Fig. 3 - 4). Such shielding is massive and very expensive.

When recording from the brain, movement of the head can change the levels of flux measured from the source (generator) since the field falls off as the square of the distance of the sensing coil from the source.   This is particularly troublesome if the MEG measurements are taking place over a period of seconds because it is difficult for a patient or subject to restrain movement for that period

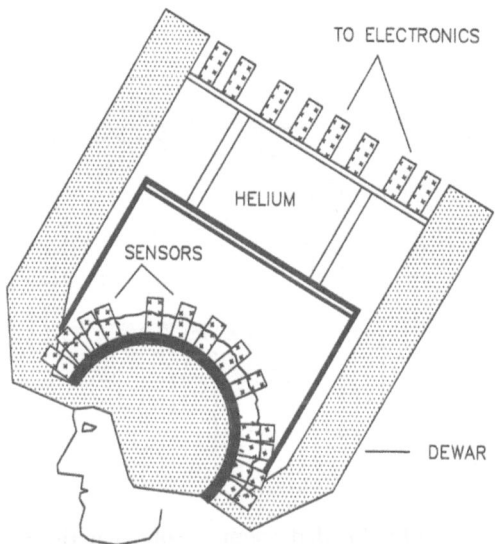

3-3: 60 Channel biomagnetic system. Schematic representation of a whole head, helmet system now under construction. The intent of the system is to measure a distribution of fields over the whole head without having to move the instrument or the subject. The advantage of this type of system is that simultaneous fields arising from the whole brain can be captured at the same time.

of time. If the subject is required to make a response, for example to move a hand or foot, small movements of the head may also occur. Two solutions have been used. One method is to fix the head in a relatively unmovable position using an adjustable vacuum cast or a similar device. This can still be uncomfortable for the patient, especially if the head has to remain in the same position for many minutes. The second solution is to keep track of the movement of the head relative to the sensing system. If the relative position of the head is known, the absolute flux values can be adjusted according to the distance of the sensing system from the head.

Magnetic fields which were clearly the result of electrical activity in the brain was first recorded by Cohen (1972). He observed spontaneous rhythms in the MEG which were in the alpha frequency range (8-12 Hz). Following those initial observations the field of

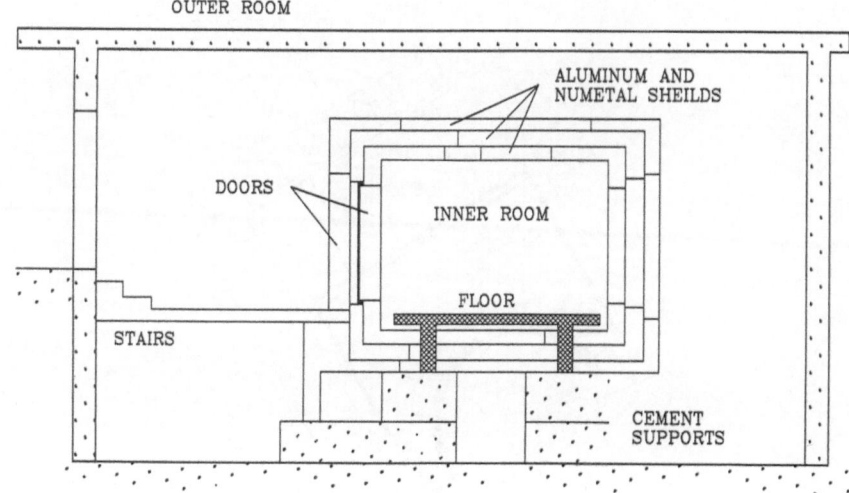

OUTER ROOM

ALUMINUM AND
NUMETAL SHEILDS

DOORS

INNER ROOM

FLOOR

STAIRS

CEMENT
SUPPORTS

3-4: Typical magnetically shielded room. Note that the room is constructed
to minimize vibration of the shielding.

magnetoencephalography has grown rapidly and several recent
reviews (Romani et al. 1983; Williamson et al. 1983; Weinberg et al.
1985; Romani and Rossini 1988) document the relevant research and
theoretical issues.

3.2 MEG MEASUREMENTS AND GENERATORS

Conventional overt behaviourial measures, such as reaction time
and percentage of correct responses are good indices of motor
capabilities but give only limited information about the processes
underlying that behaviour.   For example, a person making many
errors may have deficits in organizing motor output although a good
deal of the information upon which the output depends may have
been correctly processed. Deficits of sensory and motor function can
result in changes in latency, distribution and amplitude of evoked
potentials. However, In the past 20 years there has also been a
growing body of evidence that relates evoked electrical potentials can
index complex cognitive function.   To a large extent this evidence
consists of an analysis of components of evoked potentials, and more

recently the analysis of evoked magnetic fields. A component of an evoked potential (EP) or an evoked field (EF) is a systematic fluctuation in the field (usually but not always a reversal of polarity) defined by its latency after sensory stimulation, or its (negative) latency preceding muscle movement. When different stimuli or responses result in components, and those stimuli and responses temporally overlap, components are sometimes said to overlap. However it has been shown that the information content of different stimuli, presented simultaneously, can differentially modify a component. Two broad categories of components are discussed in the literature, those which reflect properties of stimulus energy (sensory evoked potentials), and those which index the nature of information processing within the brain, either with respect to stimulus input or response output (sometimes called event related potentials-ERP).

The analysis of the electrophysiological effects of input from the environment allow the separation of processes in the brain associated with the registration of sensory input, selective and generalized attention, memory update and storage, the interaction of the input with previously stored information, and the processes associated with retrieval and preparation for the appropriate motor output. When information input occurs, the early brainstem electrical activity occurring within about 10 msec. after the input are transmitted to the surface through volume conduction. These early components index function in different parts of the brainstem, depending to some extent on the sensory modality. Following that in the interval from approximately 10 msec. to 80 msec. the information has been transferred from brainstem to thalamus and other midbrain structures. The literature indicates that these midlatency waves index thalamic function. In the interval of approximately 80 to 300 msec., complex cortical function occurs in the initial and later stages of information processing. For example, it is generally agreed that P300, a positive potential occurring at about 300 msec. after input is an index of memory update (Fig. 3 - 5).

MEG generators

The evidence that components of evoked potentials indexed information processing led to the question of whether it was possible to determine the physiological and neuroanatomical systems responsible for those components. The logic was, and is, that the

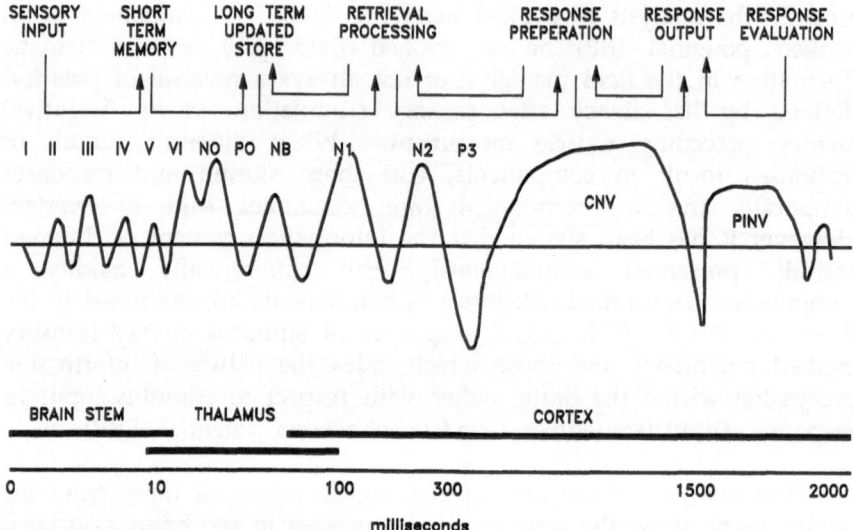

3-5: Information processing time line. When information is processed as the result of some input the information proceeds through stages of processing that are associated with evoked electrical and magnetic fields arising from different parts of the brain. Note that during some intervals feedback loops are established.

spatio-temporal systems responsible for components of evoked potentials must be involved in the processing of the information that influences these components. For example, if it were possible to locate a generator or a system of generators in the brain that are responsible for changes in the amplitude and latency of N100, then that system must be involved in processing, storing and retrieving the information, and organizing the output for selective attention. Furthermore, generators (or generator systems) could theoretically be identified for components of evoked potentials which result from pathology of the brain.

Since estimates of generators involve, in one way or another, the measurement of the electrical activity of the generator, MEG was seen to have a significant advantage. Generators, for example a dipole generator (discussed earlier in Part 2 of this book), produce electrical fields that are measured with electrodes on the surface of the brain or scalp. These volume conducted electrical fields are modified and smeared by the resistance and capacitance properties of the brain mass, the skin and the skull. Since magnetic fields travel freely and are unchanged by brain, skull and skin, they are able to give theoretically a more accurate estimate of generators.

The concept of a generator or source as an aggregate of distributed activity is one which has not been sufficiently appreciated and is pivotal for any description of systems of the brain responsible for the processing of complex information, i.e., cognitive function. Although this has been covered in Part 2, relevant points will be reiterated here.

When considering what is meant by a generator it is important to distinguish between generators associated with cognitive function, i.e., information processing which accesses memory systems, and generators associated with those events of the brain which are the first stages of sensory processing (and indeed also the last stages of motor output). The distinction between sensory and cognitive processes imply a differentiation between those brain functions that code the so-called physical characteristics of input (i.e., stimuli) and those processes which extract the meaning of stimuli. The meaning of a stimulus is defined by the demand characteristics of the stimulus (what happens as the result of stimulation). For example, an auditory stimulus which changes in frequency or intensity produces changes in the brain primarily attributable to registration processes: the initial sensory processes that are the first link in a long chain of the effects of that stimulus on the brain. That same stimulus has cognitive value when it is given demand characteristics, which requires processing that goes beyond the registration stage, when many parts of the brain are interacting.

It is frequently not clear what is meant by the "localization" of a generator or system. The reason this is important is that MEG research, and also EEG research, is very much in the business of localizing sources. The problem can be reduced by making a distinction between localisation and location. Localisation usually implies the generator is localised within some small site or space, as distinct from distributed. The term location does not prejudge the data, so the generator could be local or distributed. A distributed system may be usefully described as a constellation of current dipoles, distributed in some three dimensional space, each having different vectors and each resulting from some excitatory and inhibitory processes.

There are four types of generators that can be distinguished: (a) localized and stationary, (b) localized and non-stationary, (c) distributed and stationary, and (d) distributed and non-stationary. "Localized" means that the generator can be modelled as a single dipole. When a dipole is stationary it does not change its location, given that it is measured at the same time relative to successive (stimulus) inputs. Most approaches to the analysis of generators in MEG (and EEG) assume that the generator of a component is stationary.

This assumption is frequently not realistic, especially when the interval of the component is long, e.g., 100 msec. It is unlikely that only a single, highly localized generator would be active over such a long period of time. Furthermore, the inverse solution is not unique, i.e., several different combinations could result in the same equivalent dipole.   Fig. 2 - 12   is a simple example of how different combinations of dipoles could add to the same equivalent dipole.

Equivalent dipoles are theoretical descriptions of a group of dipoles, a "vector average". The computation of equivalent dipoles can sometimes be misleading. For example, a solution which places the equivalent dipole for auditory stimulation in the centre of the head, because the computed dipole is a vector sum of single symmetrical dipoles in each of the two hemispheres, would be misleading. It is for this reason that studies of visual, auditory and somatic sensory systems have comprised a large body of MEG research. There is relatively well known neuroanatomy and neurophysiology associated with these sensory functions. In the case of known neurophysiology the location of generators associated with components of sensory evoked potentials can be validated with what is already known about the system of motor or sensory function.

Distributed generators are those which are located in different places and are simultaneously active. Components of ERPs are almost certainly attributable to distributed, non-stationary generators. If a component of an ERP is one which develops over a period of as much as 50 msec. it is quite likely that, for windows within that interval of time, multiple generators are involved which are spatially distributed and simultaneous active, and which are changing over the interval of the defined component, i.e., they are non-stationary.

Regional generators

The concept of a regional generator is that there exists a defined three dimensional region of activity responsible for the distribution of voltages observed over some unit of time, which may be defined as the entire EP or some subset, i.e., some predefined component of the EP. The brain is always active in respect to the processing of information and the region of activity is always defined in respect to some defined period (in the EEG) of interest, whether or not the period is time locked to stimulus input or to response output. For spontaneous activity a region of interest may arbitrarily begin at some time of observation and continue over an arbitrarily defined period. An estimated region is the anatomical distribution of generators overlapping in time, during the defined interval of interest. Equivalent dipoles (or sources), and regions of activity, can be thought of as evolving or changing over time, i.e., growing and waning or changing their configuration.

It is almost certain that widely distributed systems are involved in complex information processing. If ERPs or components of them are known to index complex memory processes, and particularly if they are modality non-specific, it is very likely that a complex neurophysiological system underlies the generation of that ERP component.

The inverse problem for MEG

Accurate location of generators is theoretically made possible because magnetic fields are not attenuated or distorted by the skull or other tissues of the head. Methods of estimating generator locations make use of the simplifying assumption that neural generators can be modeled as current dipoles and that the head can be approximated as a sphere. The implication of this assumption is that the component of the magnetic field which is radial to the surface of the sphere, i.e., the skull, is due to the primary current flow of a dipole tangential to the skull (Fig. 3 - 6). Physical models have been used to study the problem of generator location in which current dipoles were implanted in known positions within a human skull, in vitro. A least-squares iterative method was used to find the

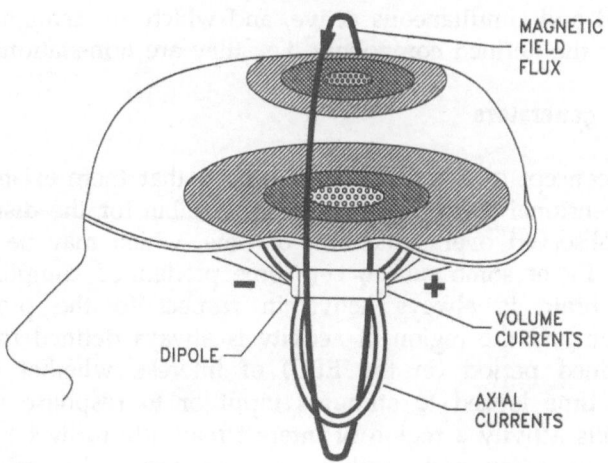

3-6:   A theoretical equivalent dipole.   The theoretical equivalent dipole is actually modeled as a point-source.   The magnetic field flux that is measured is orthogonal to the orientation of the dipole.   An analysis of the distribution of the fields allows an estimate to be made of the location and orientation of the dipole.   The advantage of MEG is that the fields are not distorted by the skin, skull and brain.

parameters of a dipole such that the sum of squared differences between the recorded and predicted data was minimized.   The smallest three dimensional error in the location of single dipoles was shown to be 3.5 mm using this procedure (Weinberg et al. 1986).

The direction of the field measured outside the head can be determined from the configuration of in-going and out-going fields. The right-hand rule can be applied to determine the polarity of the dipole producing the field. If one's thumb is pointed in the direction of the positive pole then the direction of the field follows the curl of the fingers of that hand.

Several attempts have been made to use physical and computer modelling of the inverse solution. An important question arising when estimating the location of current generators is how to accommodate the actual shape of the head. In order for there to be accurate estimates of the location of a dipole with respect to the origin of a sphere fitted to the surface of the head, the head shape must be digitized. Digitizing the head shape allows the computation of the distance from each point on the surface of the head from the centre of a theoretical sphere. The field measured from each gradiometer position is then adjusted to what they would be if they were measured on the surface of the sphere. Those adjusted values are then used in the inverse solution. Part 2 described this approach with EEG sources.

Two methods have been used to estimate the location of sources. One method is derived directly from the known physical relationship between the distance between extrema (maximum and minimum strength of field leaving the head) and the depth of the dipolar source producing those fields. For any source of given strength the further away it is from the centre of the sphere, the greater will be the distance between extrema. Therefore the distance between extrema is a measure of depth of the dipole. If a sphere is assumed then only dipoles tangential to the surface of the sphere are seen. There are no fields from radial dipoles, and no fields from a dipole located exactly in the centre of the sphere. Since the dipole orientation is orthogonal (90 degrees) to a straight line between the extrema of the dipolar fields, the orientation of the dipole can be determined by drawing a straight line between the extrema. The direction (which end is positive) can be determined by which of the extrema are ingoing to the head and which are outgoing. This procedure works well if there is a dipolar field. However, in reality, many of the fields measured are not dipolar, and may contain several maxima and minima. In order to make accurate estimates in these cases a least-squares fit is used. A dipole is assumed to be in some starting position in the sphere. The logic of this procedure is straightforward and has been detailed in Part 2.

The difference between those process responsible for sensory registration and the later stages of information processing has been called exogenous and endogenous. Indeed it can be argued that the very concept of evoked potential components assumes the distinction between input, processing and output stages of brain function (see Fig. 3 - 5). Of course, different types of processing may be occurring simultaneously, or successively. When processing is occurring simultaneously in different parts of the brain, the electrical changes associated with that processing may summate both functionally and electrotonically, i.e., through volume conduction.

Studies of MEG evoked field components utilise measures of the amount of magnetic flux (corresponding to amplitude of evoked potentials), latency of field components and distribution of the fields over the head. These are the raw data that are used to decode the neurophysiological strategy used during information processing. These measures are frequently said to index the nature of the programmes and subroutines being developed and used by the brain. However the description of neurophysiological strategies, i.e., a model of the brain's software, does not necessarily have to include statements about anatomic structure or biochemical processes. Cognitive psychophysiology frequently uses this approach, that is to utilize the MEG to model the strategies used in processing different kinds of complex information. Those strategies can be described as the interaction between elements of a neurophysiological system.

3.3 SPONTANEOUS MEG RHYTHMS

Spontaneous rhythms are continuously changing and reflect activity of the brain that is not necessarily time locked to immediate stimulus input. Spontaneous alpha MEG activity has been localized to parieto-occipital areas by several studies (e.g., Williamson 1989) and thus far there have been no significant interhemispheric differences reported. Estimates of generator location for spontaneous activity is made difficult because normally signal averaging is required to increase the ratio of signal to noise.   Therefore if averaging is to be done the signal has to be time locked to some phase of alpha. This is the same technique that is used to average spontaneous interictal spike activity, and assumes that the same and underlying processes are occurring during each phase of the activity selected.

Grey Walter expressed the idea that the spontaneous EEG recorded from macro electrodes inside or outside the brain resulted from the electrotonic and functional summation of "brain waves" different in frequency, amplitude, and phase. His idea, also expressed by others, was that the frequency components of the spontaneous EEG came from different systems in the brain and reflected the function of these systems. If one believes this then EEG and MEG frequencies within the spontaneous bands can be realistically represented by the components of a Fourier analysis of the spontaneous activity. Consequently a legitimate question can be posed in respect to the generator systems responsible for those Fourier components. For example, in the case of alpha activity Fourier analysis of the spontaneous EEG can be used to identify an 8 Hz spectral "component" and an inverse solution then applied to the scalp distribution of that component. In order to do this the phase of the components computed from MEG (or EEG) at different scalp locations must be preserved. Two logical approaches to the inverse solution can be applied to Fourier components. One has been described by Lehmann and Michel (1989), and the other is to return to the time domain after a frequency analysis, and make an estimate of generators in one period of the sine wave representing one Fourier component (Weinberg et al. 1989). The latter procedure can be diagrammed in Fig. 3 - 7.

The first stage as shown in Fig. 3 - 7 is to record the distribution of fields over the head. The second stage is to do a fourier analysis of the recorded data for each position. The third stage is to select a particular frequency of interest, e.g., 8 Hz, and do the inverse computation for that frequency, retaining the phase information in the data from different locations. This procedure results in sine waves of the frequency selected and returns the data to the time domain from the frequency domain. The next stage is to select a point in the time domain for estimating the dipole position. The point selected can be any point but, if the solution is to be a dipole estimate then the point which results in the greatest number of phase reversals in the time domain distribution should be selected. That point is then mapped over the head, just as a point in an evoked field would be mapped. That map is then used to make an estimate of the location of a theoretical dipole that would be responsible for the map. The result is that the dipole is the theoretical location of the generator for that frequency of the magnetic field. In this way one can go about attempting to locate the generators of different frequencies components, or constituents, of an evoked or spontaneous field that is actually made up of many different frequency components.

If this procedure is used to study alpha activity (8-12 Hz) the generators are nicely located in bilateral striate cortex, a finding which fits well with other studies which conclude that alpha activity originates from striate cortex. There is recent evidence that different spindles of alpha may have different generators. A spindle is a brief change to the alpha frequency (8-12 Hz) from the ongoing dominant

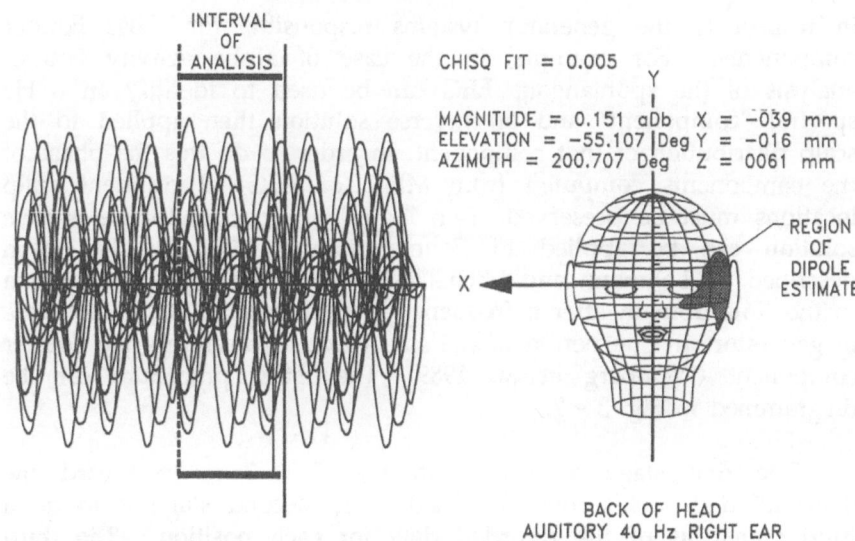

INTERVAL
OF
ANALYSIS

CHISQ FIT = 0.005

MAGNITUDE = 0.151 aDb   X = -039 mm
ELEVATION = -55.107 Deg Y = -016 mm
AZIMUTH = 200.707 Deg   Z = 0061 mm

REGION
OF
DIPOLE
ESTIMATE

BACK OF HEAD
AUDITORY 40 Hz RIGHT EAR

3-7: Analysis of frequency components. Fourier analysis is used to extract a single frequency from the MEG or EEG and plotted in the time domain, for different locations over the head (tracing shown superimposed). Dipole estimates are made for one period of the frequency and plotted as shown in a model of the head. The dipole estimates fall within a region of the brain that is active in the production of that frequency. If that frequency is characteristic of some types of information processing the region of the brain active in that processing can be identified. CHISQ FIT is a measure of the accuracy of the dipole estimate, it is the sum of the difference between the observed values and the predicted values. Elevation and azimuth define the orientation of the dipole, magnitude its strength and x, y and z its location.

frequency band of the EEG. This frequently occurs as the EEG changes during changing states of arousal. Presumably, these individual alpha spindles combine in what is normally seen as the alpha rhythm.

The same logic could be used to study other components of the MEG. One could, if one were bold enough, extend the logic further to ask whether there are generators associated with more abstracted mathematical derivations. For example a multivariate analysis of MEG is possible to extract the factors which account for a predetermined percentage of the observed variance at each location for the purpose of a principal component analysis, and then the question can be posed as to whether there are different generators associated with those factors. We have tried this with data collected preceding speech production and have localised the principal components in the left temporal lobe.

3.4 REVIEW OF MEG STUDIES

Retinotopic organisation is mapped on the striate cortex. Brenner (1981) estimated generators associated with full field, and right and left hemifield stimulation. The stimulus was a vertical grating whose luminance varied sinusoidally across an oscilloscope with a spatial frequency of 5 cycles per degree. He estimated equivalent dipoles contralateral to hemifield stimulation oriented horizontally and symmetrically, in agreement with the cruciform model. The equivalent dipole for full field stimulation was a linear sum of those estimated for the hemifield stimulation. Okada (1982) using essentially the same technique showed that the phase lag between VEP and VEF was a function of spatial frequency of the stimulus and argued that the VEP was attributable in part to a different generator than that measured by the VEF. Other studies of sinusoidally modulated gratings report generators in the rolandic fissure near the sites from which efferent fibres influence eye movements (Lounasma et al. 1985). Weinberg et al. (1985) recorded EF resulting from stereopsis of a random-dot binocular display. Their data suggested that the generator of an early component of the EF associated with fusion was located in the striate area and the generator of a later component associated with fusion in the right temporal lobe.

EEG auditory evoked potentials have been used extensively to distinguish between generators in brainstem and cortex. However MEG studies have been, almost exclusively, concerned with cortical generators, with the exception of Cohen (1985) who reported that attempts to record 6-10 msec. brainstem EF were unsuccessful. He explained the failure to observe brainstem MEG as attributable to the possible combination of excessive depth with respect to the location of his sensing coil and generators which were radial dipoles. It is however unlikely that the assumptions of a spherical model apply to generators as deep as cochlear nucleus. The question of whether MEG is capable of locating brainstem generators is still open.   Reite (1978) first detected the auditory evoked field using brief clicks. They reported fields which were maximal in central and parietal regions but did not sample the distribution of MEG fields sufficiently to establish reversals.

Hari (1984) recorded auditory evoked fields elicited by a continuous tone of 800 msec. duration. She distinguished between the generators of N100 and the slow field which immediately preceded off-set of the stimulus. Both these generators were estimated to be in the superior surface of the temporal lobe in the primary auditory cortex. She postulated that the generators were sheets of dipoles consistent with the assumption that they were produced by apical depolarization of pyramidal cells with extracellular sinks in the superficial cortex.

Romani (1982) reported responses to single frequencies that were modulated at 32 Hz. He used steady state averaging techniques in which the samples were time locked to the modulation. He recorded averaged fields resulting from 200, 600, 2000 and 5000 Hz stimulation. The generators estimated were located in the primary auditory cortex and increased in depth from 2.2 to 3.2 cm below the scalp as the frequency of the tone increased from 100 to 5000 Hz. Galambos, Makeig and Talmachoff (1981) described what they called the 40 Hz response. This steady-state response was a sinusoidal EEG following response to repetitive auditory stimulation which was of maximal amplitude at rates between 35 to 45 Hz. Galambos suggested that the response was a superimposition of thalamic mid-latency responses. Spydell et al. (1985) examined the relationship between the phase of the auditory brainstem component and the 40 Hz component of the averaged waveform, predicting that they should be different in normals and patients with brainstem lesions. He came to the conclusion that there were independent generators. He also observed that in a group of patients with unilateral temporal lobe lesions the 40 Hz response appeared unimpaired. He concluded from this that temporal cortex was not involved in the response.

However, Weinberg et al. (1987) reported a study in which both EEG and MEG data were obtained from two healthy right-handed subjects with normal hearing during 40 Hz auditory stimulation. EEG recordings were taken from a vertex electrode, MEG recordings were obtained sequentially from a wide distribution. They suggested that a "generator-system" was active during 40 Hz auditory stimulation which included bilateral temporal cortex, and as did Borda (1984), that when the brain is in a steady-state as the result of being driven by repetitive auditory stimulation it is likely that a resonance within a limited portion of the auditory pathways could account for some of the enhancement at 40 Hz stimulus rates. They have proposed that a pathway consisting of reciprocal connections between auditory cortex, medial geniculate nucleus and inferior colliculus could be the basis of a resonance resulting in an enhancement of a response to 40 Hz auditory stimulation. This interpretation is supported by what is known about the organization of the thalamocortical auditory system in the cat (Anderson et al. 1980; Imig et al. 1983).

Several MEG studies have utilised transient and steady state somatosensory stimulation because there is reasonable agreement about the anatomy of afferent somatosensory projections to the cortex. Brenner (1978), Kaufman (1981) and Okada (1981) did the early experiments in which they showed that steady-state stimulation of the median nerve produced magnetic fields that suggested dipolar generators in the contralateral somatosensory cortex.

Hari (1985) studied both median nerve and peroneal nerve stimulation in the same experiment. Two different types of magnetic responses were reported, one type (SI) with generators near the primary projection areas of somatosensory cortex (30-80 msec. and 150-180 msec. latencies) and a second type (SII) near Sylvian fissure (90-125 msec. latencies). The SI response was lateralised with respect to contralateral stimulation. The SII response occurred after both ipsilateral and contralateral stimulation. Previous studies of these two cortical sites (somatosensory cortex and Sylvian fissure) suggest that they are not hierarchically organized anatomically, or functionally, but that the Sylvian cortex functions to integrate somatosensory input with input from other sensory modalities and with memories. Magnetic field distributions differentiated between equivalent dipoles active in SI and SII whereas electrical field distributions did not (Hari et al. 1980).

In a recent study steady-state stimulation has been used to study the processing of information delivered simultaneously and in phase to two modalities (Weinberg et al. 1987). Three conditions of steady-state stimulation were examined: tactile (T); auditory (A); and simultaneous (B). Fourier analysis was used to compute amplitude and phase of the 40 Hz component from the MEG average at each recording position. These values were then plotted as vectors in polar coordinates. Since a phase difference of 180 degrees is equivalent to a polarity reversal of the same signal, the average phase was calculated as that angle for which the root-mean square amplitude of all vectors projected onto that angle was maximal. The resulting amplitude values were used to produce isocontour maps of the field over the surface of the head.

Simultaneous, in phase, auditory and tactile stimulation resulted in an EEG at the vertex that showed a following response which was similar in waveform and period to the tactile and auditory stimulation alone. The observed MEG data is shown in Fig. 3 - 8. A complex distribution of fields with patterns of maxima and minima were observed in both hemispheres. The somewhat unexpected results was that the arithmetic sum of the tactile and auditory fields is very close to the empirically observed field resulting from stimulation.

The obvious interpretation of this finding is that the result of simultaneous stimulation is the sum of what would be expected from the stimulation of each of the modalities separately, as if the brain were simply doing a linear sum during simultaneous processing. Taken as a whole, the observed bilateral fields were what would be expected if the tactile and auditory stimulation were producing fields which summed on left side, but not on the right. Since tactile stimulation does not result in ipsilateral fields, only the right hemisphere auditory fields are available to sum in the right hemisphere. If it is the case that the combined fields for dual modality stimulation are the sum of the fields for single modality stimulation it suggests that the 40 Hz response is confined to sensory systems related to the modality of stimulation; if there were an interaction of these modalities in the 40 Hz response the combined fields would undoubtedly not be a simple linear sum of the two.

From a more general perspective the data suggested that the two modalities may be driven independently by modality-specific stimulation and supports the interpretation that the 40 Hz response reflects a resonance in the sensory system stimulated. Plots of the topographical distribution of phase angles for an extracted spectral component may be used to support an assumption that more than one generator is involved. Since the brain tissue, skin and scalp are transparent to magnetic fields there should be no phase difference

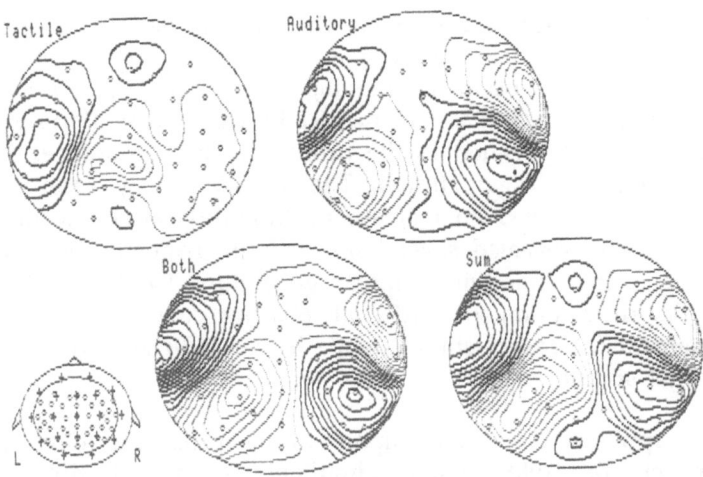

3-8:   Maps of magnetic fields resulting from auditory and tactile stimulation. The fields resulting from right forefinger tactile stimulation are lateralized and resulted in a dipole estimate for the left somatosensory cortex.   The fields resulting from auditory stimulation are bilateral and resulted in dipole estimates for right and left auditory cortex. Fields labeled both resulted from simultaneous tactile and auditory stimulation.   The fields labeled Sum are the linear sum of the Tactile and Auditory fields.   Note the similarity of the Sum with the observed fields resulting from combined stimulation of auditory and tactile systems.

over the head if there is a single generator; if only a single dipole were active, all magnetic data would be in phase; there would be no topographical distribution. The systematic change in phase from posterior to   anterior locations observed after auditory, tactile and simultaneous stimulation is consistent with the interpretation that two or more dipoles   summate to produce the sinusoidal 40 Hz field normally observed, even though the fields may appear dipolar. Therefore although the estimated locations of equivalent dipole generators are consistent with the appropriate functional cortical sites, it does not mean that the generator of the 40 Hz responses is entirely cortical.

Weinberg et al. (1987) conducted a study where simultaneous EEG and MEG responses were recorded from subjects who were engaged in a task involving perceptual discriminations of moving visual stimuli. In this study where the subjects detected and covertly counted rarely (p = 0.3) occurring moving target shapes and ignored frequently (p = 0.7) occurring moving non-target shapes, a 600 msec. ERP component that was symmetrically distributed (left-right) was identified as being related to the probability of appearance, detection and counting of Target stimuli. The simultaneously recorded MEG responses also showed an increase in amplitude at this same latency (600 msec.) and revealed several reversals in magnetic sense over the scalp suggesting underlying dipolar generators. Unlike the EEG data the scalp distribution of ERFs at 600 msec. was quite left-right asymmetrical for Target stimuli while it was left-right symmetrical for Non-target stimuli.

By applying a current dipole fitting procedure developed by Harrop et al. (1986, 1987), simultaneous current dipoles were estimated for the scalp distributions of magnetic flux 600 msec. following the appearance of Target and Non-target stimuli. The resulting generator estimates were interpreted with a human neuroanatomy atlas and were found to be located in brain regions that were consistent with current neuropsychological theory regarding cortical contributions to the performance of such a visual perceptual task.

These findings are of interest for two reasons. First, they illustrate an instance where the EEG and MEG results are apparently consistent in identifying the same component (i.e., a 600 msec. amplitude peak related to task contingencies). Second, they illustrate an instance where additional information is gained from the MEG in that the symmetrical EEG scalp distribution suggests one central generator associated with this component, while the MEG scalp distribution suggests that the 600 msec. component is associated with multiple simultaneous generators that are asymmetrically distributed in the brain.

An electrical potential which is known to index attention and information processing is the contingent negative variation (CNV), a slow increase in cortical negativity occurring between two stimuli one of which is a warning stimulus and the second of which requires a decision or response (Fig. 3 - 9).

For example, the CNV is very sensitive to distraction and is a measure of the degree to which the system (i.e., brain) can tolerate

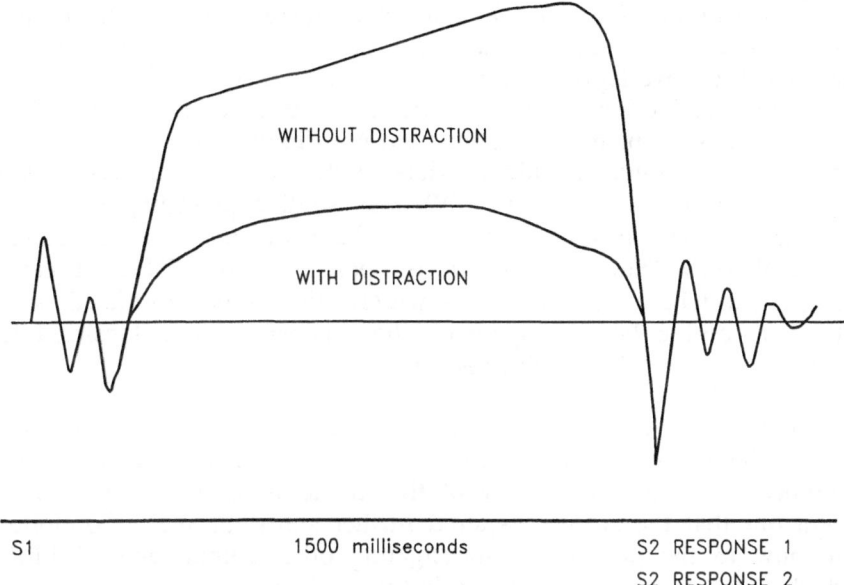

S1   1500 milliseconds   S2 RESPONSE 1
                                                                        S2 RESPONSE 2

3-9: The effect of distraction on the contingent negative variation (CNV). S1 is a ready signal, followed by S2 which signals either of two possible response. Distraction occurs in the S1 - S2 interval. By studying the dipole sources associated with the CNV those parts of the brain responding to distraction can be identified.

extraneous input in the processing and preparation of motor output. There is a good deal of evidence that N100 is an index of selective attention, i.e., the degree to which the person can shift attention from one input to another. These are only some examples of the ways in which electrical changes in the brain index the nature and state of processing, and the processing load on brain systems.

One of the first MEG reports of P300 was Weinberg (1985) which noted a similar morphology of MEG and EEG responses; however the MEG response, unlike the EEG response showed reversed polarity over posterior hemispheres, suggesting deep midline generators, although no isocontour mapping was done. Okada (1983) reported MEG data which were interpreted as evidence for the generator of P300 being located in the hippocampal formation.

Some of the issues as yet unresolved includes the nature of the distribution of generators underlying the shift, whether the shift is composed of overlapping components, each of which has different generators, and others. It is clear from all MEG studies of the CNV that the maps of field distributions are not dipolar and designation of equivalent generators should be done with caution. Weinberg (1983) reported studies in which the CNV was studied preceding the same movement that produced a Bereitschaftspotential, or "readiness potential" (BP). They reported two components of the CNV: an early component whose generator was widely distributed associated with attention, and a later component whose generator was localized to motor cortex, associated with response preparation.

Fiumara (1985) however reported data suggesting the generator of the later component of the CNV was in frontal lobes. They commented on the complexity of the distributions of the fields and point out that the subject required further study. Several laboratories are now in the process of investigating other components of ERPs which are associated with selective attention and information processing.

Movement-related electrical brain potentials were first reported by Bates (1951) who observed systematic changes in the EEG *after* hand movements. The first observation of a sustained, slow potential, shift *preceding* a motor response was reported by Walter et al. (1964). Using a reverse-time averaging technique triggered on the onset of the electromyogram, Kornhuber and Deecke (1965) were able to demonstrate that non-cued voluntary (self-paced) movements were preceded by slow negative shifts over the scalp beginning 1 to 2 seconds prior to the movement, which they referred to as BP. This was confirmed by Gilden et al. (1966) who coined the term "Motor Potential" (MP) for the cortical potential accompanying a voluntary movement. The temporal structure and topography of the BP has been extensively studied (Vaughan, Costa and Ritter 1968; Deecke, Scheid and Kornhuber 1969; Deecke et al. 1984). Accordingly, there has been a great deal of interest in the question of whether there are specific and well localized generators of these brain potentials and their possible use for the study of normal and pathological human motor control.

An important thrust in BP studies has been the search for the generators of different "components" of the electrical changes preceding, accompanying and following voluntary movements. Known neurophysiology and anatomy of motor systems suggests that systems of generators may be involved in voluntary movements and that they may be both distributed and complex. The neurophysiological systems responsible for voluntary purposeful movement involve several interacting processes which are frequently difficult to separate. Ghez (1974, 1985) has pointed out that corticospinal neurones play an important role in the control of distal muscles, and that they interact at the level of the cortex in the coding of the force to be exerted for the anticipated movements.

In addition the supplementary motor areas (SMA) play important roles in the programming of motor sequences. The premotor cortex plays a role in the preparation for sensory guidance and the parietal cortex plays an important role in the analysis of spatial information necessary for and preceding guided movements. Until recently it was generally thought that many parts of the cortex were comprised of "centres" where specific functions were encoded and/or controlled. This concept is now considered wrong. As Ghez (1985) pointed out: "local areas of cortex function as modules that transform the complex information they receive and direct their output to other modules according to specific rules. Moreover, specific cooperative arrangements exist between widely separate areas of cortex and between the cortex and subcortical structures, an arrangement Mountcastle referred to as *distributed processing*". If this view is correct it is unlikely that highly localized specific centres are exclusively responsible for specific components of the BP. Therefore more recently there has been interest in identifying the cooperating systems responsible for voluntary movement and determining the degree to which they are functioning sequentially and in parallel.

When a voluntary movement is the beginning of guided tracking, i.e., movements that depend on visual and/or proprioceptive feedback, the BP shows increased amplitude over the parietal cortex. This is partly attributable to the arousal effects of focused attention and partly to the increased processing necessary to adjust the voluntary movement. The execution of continuous performance tasks, in which the performance lasts several seconds, are also accompanied by slow electrical potential shifts preceding each change in direction of the movement and the distribution of these electrical potentials is influenced by the nature of the specific task involved.

For example, this task-specificity has been shown for guided movements requiring proprioceptive feedback (Grunevald 1978; Lang et al. 1984; McCallum et al. 1988), and for verbal cognitive tasks. Many natural voluntary movements require bilateral coordination that require the linking of motor systems. Experiments have been done in which the anticipated voluntary movement is the sequential or simultaneous bimanual finger movements with different delays between sequential movements (Lang et al. 1988). When sequential movements are initiated there is a greater fronto-central midline electrical shift, presumably due in part to SMA, suggesting that SMA is more important for temporal, rather than spatial, coordination. These and many other studies make it clear that the distribution of generators responsible for voluntary movements is greatly influenced by the nature of the anticipated movement.

One of the first observations of slow magnetic fields accompanying voluntary movements were made by Deecke, et al. (1982) who described a magnetic correlate of the readiness potential which could be observed over the motor areas of the cortex during self-paced movements of the index finger. They termed these slow, low amplitude (50-100 fT) shifts in magnetic flux over the temporal and central scalp a Bereitschaftsmagnetfeld or "readiness field" (RF) in recognition of its similarity in form and time course to the electrically recorded readiness potential.

However, the topography of the RF also shows a reversal in direction over the region of the sensorimotor area of the cortex with outgoing field maxima laterally and ingoing field maxima medially suggestive of localized generators in the region of the central sulcus, just preceding movement (Fig. 3 - 10). Slow magnetic fields have been subsequently reported for foot and toe movements (Deecke et al. 1983; Hari et al. 1983), and speech (Weinberg et al. 1983). Although these studies usually consisted of limited numbers of recording locations over the scalp, they indicated slight differences in topography of the readiness fields for different kinds of movements. These studies employed simple movements usually consisting of brisk flexion of the limb or joint, although one study used a more difficult finger tapping task, in order to detect the contribution of pre-motor structures (such as the supplementary motor area) in a complex motor task (Deecke et al. 1985).

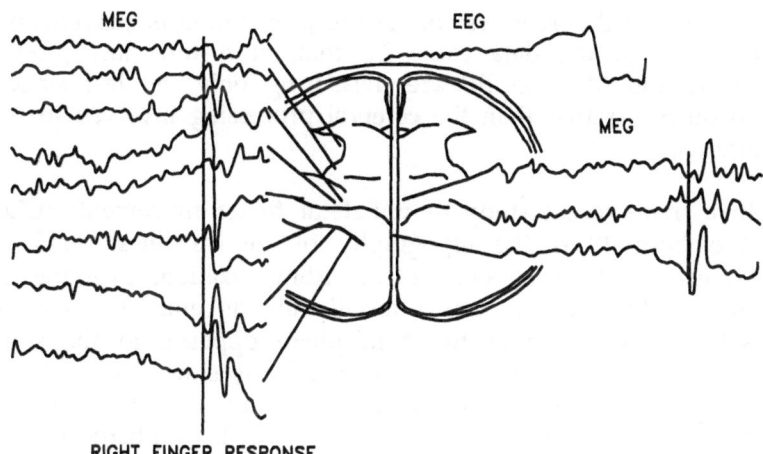

MEG   EEG

MEG

RIGHT FINGER RESPONSE

3-10:   Typical MEG response preceding finger movement.   Right finger movement occurs at the time of the vertical cursor.   Note that reversals occur in the left hemisphere but not in the right hemisphere.   This indicates that the sources controlling right finger movement are in the left hemisphere.

In these initial studies, generators of the RF were located primarily over the rolandic fissure contralateral to the side of movement, although slow fields were also noted over the ipsilateral hemisphere indicating a more detailed distribution of activity during the movement foreperiod than had been evident in EEG recordings. Additionally, large responses could be observed at the time of movement onset, although these deviations were difficult to discriminate from movement-induced artifact produced by the physical displacement of the head during movements of the limbs. Respiration-locked head movements are an additional generator of noise while recording slow magnetic fields, although this can be minimized by having the subjects arrest their breathing during the movement.

In electrical recordings the BP is not lateralized to contralateral hemisphere; rather it is widely distributed although there is a "preponderance" over contralateral hemisphere. Thus the initial MEG recordings tended to suggest that parallel processing by distributed "generators" were not involved within 20 msec. preceding movement. Since known neurophysiology (discussed above) suggests the modular contributions of different systems to the preparation of movement one interpretation of the data could be that, at a time just preceding movement, the MF and BP are measuring only the last stages of cortical output (rather than the cerebral processing involved in motor preparation).

In a more recent study of unilateral finger movements (Cheyne and Weinberg 1989) the topography of the readiness fields was studied using a large selection of recording positions over the head. Readiness fields were observed in all subjects and also showed a reversal of direction over the hemisphere opposite to the side of movement. In addition to the contralateral RF, slow fields were also observed over the ipsilateral hemisphere. Fig. 3 - 11   shows an example of RF waveforms recorded over both hemispheres while the subject performed either left or right index finger flexion.

Topographic mapping of these fields indicated that these ipsilateral slow shifts were separable from those over the contralateral hemisphere which form a localized reversal over the rolandic area in the vicinity of electrode placement C3 which lies roughly over the hand representation area of the primary motor cortex. Readiness fields begin over both hemispheres about 1 sec. prior to movement and reverse in their direction from lateral to medial locations.

However, the dipolar pattern over the rolandic cortex did not become clear in the topographical maps until about 500 msec. prior to the movement. This time period also corresponds to the beginning of lateralization of the readiness potential and also to the onset of pre-movement single-unit activity in motor cortex reported in monkeys and cats (Evarts and Tanji 1976; Neafsey, Hull and Buchwald 1978). Whether the earlier onset of the RF at lateral sites indicates that generators in primary motor cortex become active even earlier than this, or that these fields are due to other pre-movement generators needs to be studied further.

Although the term "motor field" might be applied to MEG shifts immediately prior to movement, the fields overlying the motor areas show only an increase in amplitude but no shift in polarity or orientation over the pre-movement period. Thus, distinctions between readiness and motor fields may need to take into account differences in both amplitude and topography and may require a different interpretation than that of RP and motor potential.   An MEG corre-

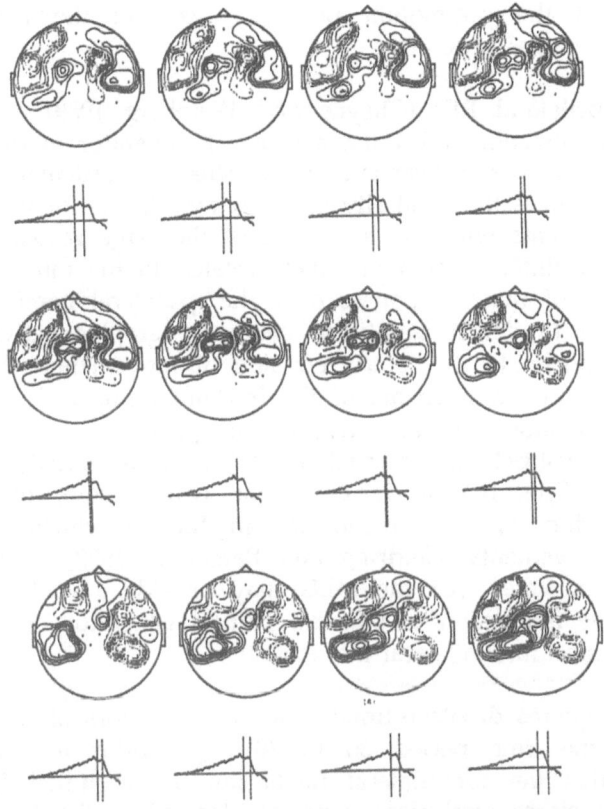

3-11:   Magnetic fields preceding and following patterned finger movements. Solid cursor indicates time of movements.   Dotted line indicates time of analysis for map shown.   Note the increased complexity as the time of analysis gets closer to the movement and the change in patterns following finger movements.   Dipole analysis of these fields suggested that feedback from sensory system is interacting with the motor systems in the preparation and performance of complex movements.

late of the pre-motion positivity is not evident in recordings over the motor cortex and has not yet been described in studies of movement-related magnetic fields.   If the MEG is measuring fields resulting from generators near the surface, where the head can be most readily modeled as a sphere, then tangential generators would be primarily responsible for the observed fields.   If this interpretation were accepted the absence of pre-motion shifts in the MEG could suggest that the pre-motion positivity seen in the EEG is due primarily to radial generators.

The ipsilateral RFs (Cheyne and Weinberg 1989) also showed reversals in direction indicating a separate generator in the *ipsilateral* hemisphere prior to movement onset. What is particular striking in the data shown is the similarity between waveforms for left and right finger movements with the exception of the large responses (fields) evident after EMG onset, most likely related to the movement itself. The absence of these responses over the ipsilateral hemisphere, and the inability to detect significant electromyographic activity in the forearm opposite to the instructed side of movement (which might indicate mirrored movements of the ipsilateral fingers) suggests that separate generators become active in the ipsilateral hemisphere even when the subject is instructed to perform strictly unilateral movements. This finding is consistent with reports of ipsilateral activity during subdural recordings in humans during instructed unilateral movements (Goldring and Ratcheson 1972; Neshige et al. 1988) and indicates that the MEG may provide a useful means of studying the interaction of bilateral cortical generators active during the preparation for unilateral movements.

Least-squares dipole estimates of the generators of the readiness field (for the time period 30 to 70 msec. prior to EMG onset) indicated that the contralateral fields can be accounted for by the fitting of a single equivalent generator located in the region of the contralateral sensorimotor cortex in all subjects. This finding supports the hypothesis of activation of generators in contralateral primary motor cortex and ipsilateral sensorimotor cortex. However, the degree to which single dipole fits accounted for variance in the data across subjects, indicates that a single dipole model is inadequate to describe all the generators active. It is probable that the generators of the ipsilateral fields are complex or non-dipolar in nature. One possible hypothesis is that ipsilateral activity may involve widespread inhibitory input to the ipsilateral motor cortex which involves multiple generators. More sophisticated modelling procedures may be necessary in order to identify the nature and extent of ipsilateral generators.

Another interesting observation of MEF is that, in some subjects, there is an overlapping of the MEF with the continued presence of generator activity responsible for pre-movement generators. Subtraction of the RF (50 msec. prior to EMG onset) from the post-EMG onset field (90-130 msec. following EMG onset) results in very clear dipolar patterns to which dipole generator estimates could be fitted to expected anatomical locations. There is probably some degree of temporal overlap of different generators in sensorimotor cortex during movement onset. This overlap certainly contributes to the variability of MEG fields observed across subjects during the period preceding and immediately following movements.

Such temporal overlapping of sensorimotor cortex generators during the post-EMG interval has also been suggested on the basis of subdural recordings (Neshige et al. 1988). Their data suggest also that variations in dipole orientation for primary motor cortex, across individuals, could also be involved. In order to clarify this issue, more data is needed to clearly identify the generators responsible for movement-evoked magnetic activity.

3.5 CONCERNS AND OUTLOOK

Although a great deal has been learned from the study of scalp-recorded movement-related potentials, it remains unclear what are the exact origin of the various components described above, mainly as a result of the limited spatial resolution of the EEG and the consequent difficulty in predicting generators from the surface distribution of these components. The introduction of the MEG and dipole localization methods based on neuromagnetic recordings has provided a new means by which to study the cortical activation during movement in humans. The study of movement-related magnetic fields of the brain is still at an early stage of development. However, the data reviewed to date indicates that the MEG offers a promising means by which to study non-invasively cortical motor function. With regard to the data reviewed here, the following conclusions can be made:

Slow "readiness" magnetic fields can be recorded prior to a variety of voluntary movements and display a topography which indicates the activation of bilateral generators, even if the instructed movement is unilateral. Generators in the contralateral hemisphere appear as early 0.5 sec before the movement and appear to be localized in the sensorimotor cortex. Consequently, the assumption of a contralateral generator being the only or primary generator of the readiness potential, based on EEG data, must be tempered.

A large amplitude "movement-evoked field" (occurring at a post-EMG onset latency of about 110 msec. for finger movements) is probably the counterpart of the MP and appears to be the result of a dipolar generator localized to the contralateral sensorimotor area. This generator is probably the first signs of movement reafferent input to cortex.

Variability in the movement-evoked field across individuals, which are much more evident in MEG than in EEG, may reflect the summation of multiple generators active in the region of the sensorimotor cortex during movement onset (i.e., both pre-and post-central generators). In some instances, it may be possible to extract simpler elements of these complex generators based on assumptions of temporal overlapping of pre-movement and movement-evoked activity.

The primary use of MEG in the study of brain pathology has been to investigate generators of focal epilepsy.

Barth, Sutherling, Engel and Beatty (1982 and 1984), comprise a team who have studied the use of MEG as a tool for the localization of depth, orientation and polarity of currents underlying paroxysmal discharges. Their data suggests that multiple generators having a defined temporal and spatial organization are responsible for different components of interictal electrical discharges. This group of researchers have also studied penicillin induced focal epilepsy in rats and reported slow field shifts associated with the development of seizures (Barth et al. 1984). This finding has been corroborated in human epilepsy. Vieth and colleagues (Vieth et al. 1988) have reported slow shifts associated with ictal activity in human patients, and perhaps even more interesting, slow shifts accompanying interictal discharges in patients with refractory epilepsy. Such interictal shifts are not seen in patients whose seizures are pharmacologically controlled, nor are they detected in normal resting subjects (Vieth et al. 1988). Slow electrical shifts during seizures, sometimes called D.C. shifts, have always been of interest, but technical problems related to electrode-scalp contact have limited the development of this line of inquiry. MEG recording, which does not require contact of the recording device with the scalp, provides a means for collecting these data without the difficulties inherent in electrical recordings.

Ricci (1985) has also studied MEG activity associated with epileptiform discharges. He has reported that the equivalent generator identified in one patient was verified by CT scan and surgery in that the localization predicted with MEG data was consistent with the location of a lesion in the structural image. Confirmation of the accuracy of MEG localization in epilepsy has come from our own laboratory (Crisp 1986; Weinberg et al. 1987). By plotting the MEG estimated dipole position on the appropriate MRI "slice" we have shown that the location predicted by MEG corresponds to lesions imaged on MRI scans in two patients.

Several other laboratories are now studying epilepsy in an attempt to identify the distribution of generators associated with both interictal and ictal states (Angianakis et al. 1988; Cohen et al. 1988). If MEG data can be successfully combined with EEG and CT or MRI, as many laboratories are now attempting, the MEG technique could make a significant contribution to the non-invasive identification of pathology underlying epilepsy in individual patients.

A further contribution of MEG to the investigation of epilepsy is the observation, first reported in some of the earlier epilepsy studies, that MEG can detect abnormal activity not present in the scalp electrical recording (Modena et al. 1982; Ricci 1983). Conversely, it has also been reported that in one patient with bilateral epileptiform EEG discharges, organized MEG fields were only present over one hemisphere (Crisp 1986). In this case, MEG fields associated with interictal discharges were clearly of a 'dipole configuration' over the right hemisphere (the hemisphere in which a structural lesion was imaged on MRI), but when the left hemisphere discharges were investigated, no distinct pattern of magnetic activity emerged.

This finding raises interesting questions about whether the left hemisphere electrical discharges were arising from a radial dipole which would not be computed with the MEG (using assumptions of sphericity of the head), or whether these signals were purely volume conducted from the other hemisphere.

MEG as an imaging device has the theoretical advantage of being non-invasive with both good temporal and spatial resolution. These advantages are not now combined in any other single imaging device. However, the combination of data from existing imaging devices into a single dynamic image would be an important breakthrough in the understanding of both normal and pathological brain function. The concept of a dynamic image is important. The future will see the development of methods of estimating distributed generators through a combination of mathematics and known anatomy and physiology. When that happens a dynamic image of changing current patterns in the brain associated with memory and complex information processing will be possible. When that happens science will have documented Sir Charles Sherrington's vision of the brain as "...an enchanted loom where millions of flashing shuttles weave a dissolving pattern, always a meaningful pattern but never an abiding one".

PART 4

STATISTICAL APPROACHES

4.1 INTRODUCTION

As for any other bio-medical measurement, topographic maps are subject to variations, not only among subjects, but even at different times within the same subject. Therefore statistical methods must be employed in order to separate the consistent part of the features of interest from the variability due to extraneous factors. For instance, in order to study and understand the significance of a spike topography, one must be able to identify the salient characteristics of that structure and separate them from minor and irrelevant variations in space, time and magnitude (alpha, muscle artifacts etc.).

A distinguishing feature of brain topography with respect to statistical analyses is its need for "multivariate" procedures. While in most scientific research variables are analyzed one at a time, the nature of brain topography requires that several variables be measured simultaneously. Multichannel readings and measures repeated in time introduce features of correlation, both temporal and spatial, that must be considered in a statistical analysis. Univariate analysis does find its place here as well, since some features obtainable through a map deserve separate study. Moreover many terms and problems of univariate statistics extend to the multivariate case. For instance one has the basic problems of distinguishing between continuous variables (usually coming from measurements) and discrete variables (usually counts); or between ordinal variables (when the values have a natural order) and nominal variables (when the values have no meaningful order); or between estimation (assessing the magnitude of a variable) and testing (verifying a certain statement about variables); and one runs into sample size problems etc.

While statistical methods have become popular in EEG and their use widespread, caution must be exercised when using them, since practically all such methods make a number of theoretical assumptions that must be met. All address certain very specific questions that may not be the most appropriate for the problem at hand. These aspects will be discussed in the next chapter.

Statistical methods of interest in brain topography address one of the following three basic problems:

a)   Estimation: Identifying the average topography associated with a certain state and its degree of variability. Two types of problems fall into this category: 1) determining the average map characteristic of a state in a given subject; 2) search for "normative data" characteristic of a group of subjects sharing the same condition, including "normality". These problems are discussed in Chapter 4.3.   An example could be the estimation of the average peak topography for a rolandic spike in a given patient and in all patients with benign rolandic epilepsy (Wong et al. 1989).

b)   Hypothesis testing: Determining whether map features recorded from two clinically different sets of patients are statistically different. This includes analysis of the effect of a treatment or of the comparison between two treatments (Lehmann 1959). An example could be the comparison between maps depicting the response to a visual stimulus in normal subjects and patients with cortical visual impairment (Bencivenga et al. 1989).

c)   Classification: Selecting which combination of the variables recorded is best able to discriminate between two or more groups of subjects presenting with different clinical pictures. This problem is also known, in other fields, as "feature selection", "pattern recognition" or other descriptors (Krishnaiah and Kanal 1982). The problem of diagnosis falls into this category, since one looks for features that can classify a patient as belonging to a certain diagnostic group with minimum error. Example: which feature(s) of a set of neurophysiological measurements can best identify meningitis patients at risk for permanent neurological deficits (Pike et al. 1990)?

In Chapters 4.3 to 4.5 we shall give an overview of the methods available to find answers to these problems, but it is important from the very beginning to realize that these three problems address different questions and require different statistical methods. It is all too easy to fall into the trap of using a method of analysis suited for one problem in order to give an answer to a different one.

The last chapter (Chapter 4.6) will address the problem of how to carry out the transition from the results of the analysis to making practical conclusions. Many experiments are conducted on a small number of subjects, use multiple comparisons and look at different questions simultaneously. From a strictly technical point of view this makes many conclusions invalid because of lack of test "power" (Larsen and Marx 1981 sec. 6.3). Yet these experiments do provide useful information, especially considering the scarcity of material and of resources available to many researchers. We shall discuss the problem and make some suggestion to render these experiments useful but not misleading.

Detailed discussions of the topics mentioned in this part may be found in the references cited throughout it as well as in the technical monographs by Mood et al. on mathematical statistics (1974), by Lehmann on estimation (1983) and testing (1959) and in the three volume compendium by Kendall and Stuart (1977, 1983). Also, good introductions to techniques specific to multivariate statistical analysis are Gnanadesikan (1977) and Dillon and Goldstein (1984).

4.2 ASSUMPTIONS AND DIFFICULTIES

Any statistical method of analysis is based on a set of theoretical assumptions; these often translate into practical limitations on the type of data that can be submitted to that analysis and on the type of questions that can be addressed. Some limitations creep into the experiment in ways which are not under the control of the investigator, thus creating intrinsic difficulties for the analysis. In this chapter we shall discuss a few of these assumptions and difficulties as they apply to the study of topographic maps.

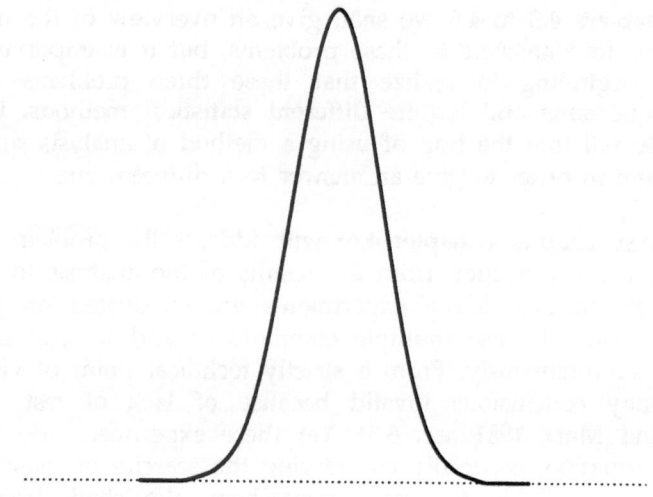

4-1:   Normal distribution. The abscissa is the value of the variable, while the ordinate is the frequency of occurence.

By far the most common assumption required by traditional statistical procedures is that the variables under analysis have a "normal distribution" (Gaussian, Larsen and Marx 1981 Chapter 7). This means that the frequency histogram of such variables must follow a precise mathematical formula (ibid., sec 7.1) which in the univariate case is represented by the "bell shape" so often mentioned throughout the literature (Fig. 4 - 1). In the multivariate case one gets a higher dimensional "bell" which is usually not easy to display (see Fig. 4 - 2 for the simple two-dimensional case). Some stringent conditions required by the "normal" distribution cannot be met by any variable coming from concrete observations. For instance, the "normal" distribution requires:

- a continuous variable, ruling out all forms of discrete and qualitative variables
- an absolute symmetry in the distribution, seldom present in practical variables that tend to be skewed
- all magnitudes, both positive and negative, to be possible; this can generate absurdities when one talks, for instance, of frequencies or skull sizes.

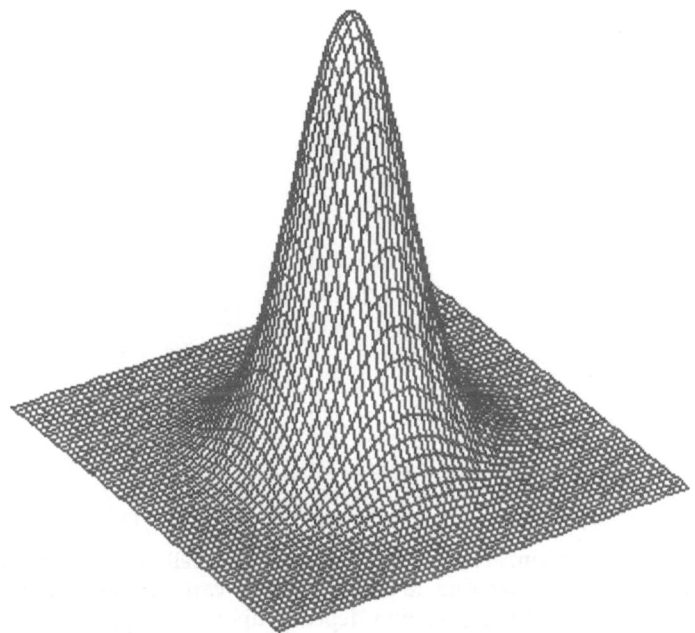

4-2: A bivariate (or 3 dimensional) normal distribution.

The assumption of normality can often be maintained, however, as a convenient approximation, thanks to a powerful mathematical result known as the "central limit theorem" (Larsen and Marx 1981 sec 7.4). This states that under very mild conditions the distribution of a variable is approximately normal if a large number of factors add up to generate the value of that variable. Thus the assumption of "approximate" normality can be legitimate, but the underlying approximations must be kept in mind when drawing conclusions from the results of statistical tests. In particular the observed frequency distribution of a variable should be analyzed to rule out outstanding features of non-normality. This can be done either by using a "goodness-of-fit" test (ibid., sec 9.3) or, at least by visually analyzing the histogram (Fig. 4 - 3). Note, however, that there are non-normal distributions that have a bell-shaped histogram but do not enjoy the theoretical properties of the normal distribution. An extreme case is given by the Cauchy distribution (Mood et al. 1974, sec 3.5), which can be obtained as the quotient of two independent

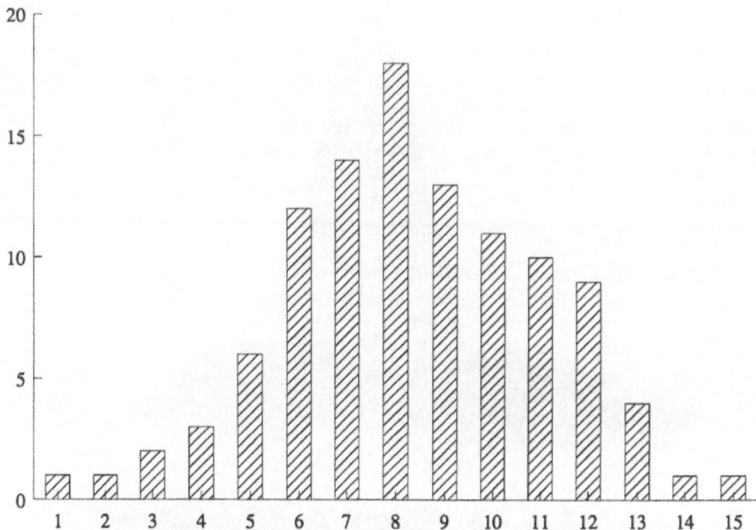

4-3:   A bar histogram. Notice the approximate bell shape with a little asymmetry. This could be due to either sample variation or non-normality. Care should be taken if interpretation depends upon assumption of normality.

normal variables but is frequently used as a counter-example to almost any good property of the normal distribution.

Tests often make assumptions on the degree of correlation among the variables under study. Most univariate methods require "statistical independence" (Larsen and Marx sec 2.8), while multivariate tests rely on the use of the "variance-covariance" matrix (Morrison 1976 sec 3.2), which takes correlations into account in the proper way.   While the concept of independence has a precise mathematical definition, it is sufficient, for practical purposes, to rely on the common meaning of the word to detect its presence in a statistical analysis. Certainly neighbouring channels recorded simultaneously are not independent and would require the use of the variance-covariance matrix to quantitatively define such dependence. Fig. 4 - 4 shows different ways that 2 variables x and y can be mutually correlated. Can we say that two consecutive spikes from the same subject are independent? Are two spikes ten minutes apart independent? Are spikes from siblings independent? The effect of dependence of data can be unpredictable and   little is known about it in general. As a result the

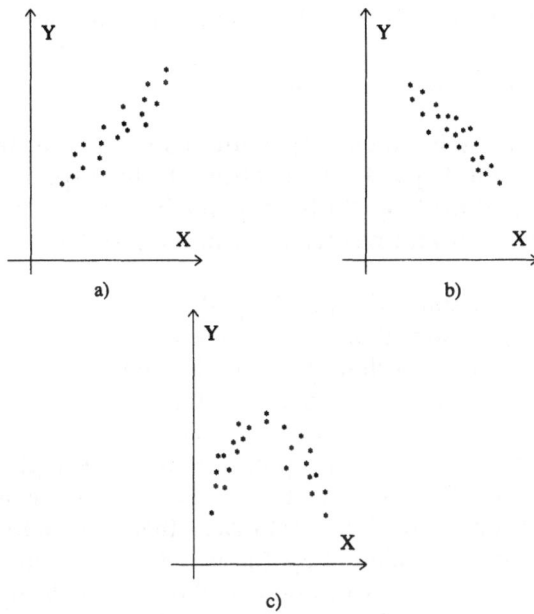

4-4:   Three type of correlated data. In a) X and Y grow together: positive correlation. In b) they grow in different directions: negative correlation. In c) there is a relation between X and Y, but it is not linear. In this case the correlation coefficient would be low, hiding the close association between the two.

best course of action is to eliminate it, to whatever reasonable extent, in the design phase of the study.

A special type of dependence that is particularly relevant for brain mapping is time dependence. Maps recorded a few milliseconds apart tend to be similar and hence it may not be proper to analyze consecutive maps of, say, an evoked response, as independent. Time dependence, however, has been successfully used to analyze temporal structures of physiological patterns. In particular, Fourier analysis, autocorrelations, cross-correlations and information theory have been tested as tools for detecting physiological and pathological features of mapping (Gevins 1989; Lopes da Silva 1989).

Other technical requirements exist that apply to specific tests, but their discussion would go beyond the scope of this book, so the reader is warned to ensure that they are satisfied whenever a statistical test is used or that the ensuing approximation does not invalidate the conclusions of the test.

Difficulties exist in connection with statistical tests that are more practical and often beyond the control of the experimenter. When planning an experiment or studying a published report these should always be considered and resolved whenever possible.

The most common difficulty is perhaps the problem of sample size. It is widely known that when collecting data it is preferable to have a large sample. Technically this is due to the fact that the variability of the statistics obtained through a random sample decreases as the sample size increases. For instance the variance of the sample mean is inversely proportional to the sample size (Larsen and Marx 1981, sec 7.3. See also Fig. 4 - 5). In other words by using a larger sample one can obtain estimates that are more likely to be close to the searched value, hypothesis tests more likely to detect existing differences and classification methods more likely to produce correct classifications.   Techniques exist for deciding what sample size is optimal in a given situation and these are usually discussed in statistics books just following the method in question. The actual decision, however, is often dependent more on financial, temporal or logistic problems than on theoretical considerations. This is a particularly important problem when one deals with rare diseases or even with healthy populations, since it is sometime difficult to convince healthy people to submit to EEG testing (see chapter on normative data). A good practice is to report the degree of precision associated to the sample size that one has been able to use: standard deviation for estimates, "power" (which is the probability of detecting a difference of a given magnitude) for hypothesis testing, or a similarly appropriate measure.

Another common difficulty lies in the fact that when selecting a sample for an experiment, one assumes that it is drawn from the desired population, i.e., that all subjects are similar with respect to the condition under analysis and any other related factor. If this is not true the results could be misleading, in that they could relate to a population other than that intended. This, however, is usually impossible to control. Subjects may have medical conditions unknown to all, factors considered irrelevant for the analysis may in fact affect the cerebral activities under study, the condition under study may

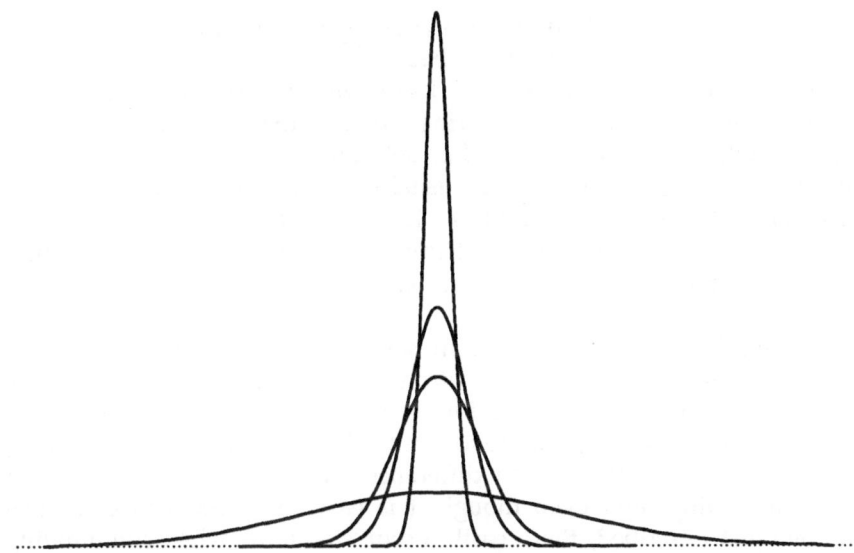

4-5:   Distributions of the sample means based on samples of different sizes. The broadest curve is the distribution of a single observation, with a large scatter. The other three, in decreasing order of variability, are from samples of sizes 10, 20 and 100 respectively. Notice how the large samples are most likely to fall close to the desired population mean.

consist of several different variants quite different with respect to the feature under investigation. There is no remedy against this problem, except trying to identify the population of interest in the best possible way and keeping always a dose of healthy reservations as to the presence of extraneous but influential factors.

When drawing a sample one also assumes this to be done "at random". Much has been written about what this means and several sampling schemes have been developed to obtain the most informative data for a certain problem (Cochran 1977). The purpose of random sampling is twofold: a) avoid the creation of a systematic bias in the sample, which is equivalent to sampling from a population other than the one intended, b) correctly use a mathematical framework for the probability calculations needed in the study. In other words if a sample is not done at random one may not be able to say anything about the variables under investigation.

In practice, sampling is done by including in a study all the patients with the same disease who have been tested in the laboratory during a given time period. This is a legitimate method, as long as no sources of bias develop during that time. A slightly more questionable issue concerns the selection of samples from the "healthy" population, since this usually consists of members of the hospital staff and their families. There are no general reasons to imply that this sampling method introduces any bias, but its validity must be assessed on a case-by-case basis.

A special case of non-random sampling occurs when a variable not considered in the analysis strongly influences the observations. For instance, if a group of healthy patients is compared to a clinical group of different age, differences may be detected that are due to age rather than to the clinical difference. This phenomenon is known as "confounding" and even though it is easy to detect it in cases like the one just described, there easily occur situations when confounding goes unnoticed and may create misleading conclusions.

As mentioned earlier, a major difficulty is provided by the multiple comparison problem. Because of the presence of variability, the conclusion of every statistical test can be guaranteed only up to a certain degree. Sampling itself generates the possibility of erroneous conclusions: what if, just by chance, we randomly select within a healthy population, a sample of individuals all having a thicker skull? Then the corresponding data would be somewhat non-representative of healthy individuals. This residual degree of uncertainty is controlled by stating a "confidence interval" (Larsen and Marx 1981, sec 5.9) for estimation, or a "p-value" (ibid., Chapter 6) for hypothesis testing. But once this is done, what happens if we analyze a large number of variables in the same experiment, either simultaneously or separately? Then of course the probability of picking up spurious results due only to unfortunate sampling increases rapidly. For instance if our method of estimation is correct 19 times out of 20, we analyze 20 variables and one of them gives a "significant result" and makes everybody excited, it is possible that variable corresponds to that one mistake predicted, on average, by the theory.

When analyzing topographic datasets this problem of multi-dimensionality is ever present and dangerous. One could consider the following examples: a map consists of several readings, one for each electrode;   usually one considers a sequence of maps, rather than a single one; each map may contain several features of interest; one may want to conduct several tests on the same set of data because of lack of subjects. The probability of obtaining spurious result in such situation can become enormous. Methods for avoiding erroneous conclusions from it and to extract valid statistical information will be discussed in Chapter 4.6.

Finally, the most common and treacherous practical problem is that of correct data collection. The results of a statistical test are only as valid and reliable as the quality of the data used for it. The expression "garbage in, garbage out" is very appropriate indeed. Many can be the sources of "dirty" data in general and even more so in the analysis of maps. For instance, machine setting, electrode placement, artifacts, manually entered data, inaccurate dictation, file editing and others can generate a large number of errors. No amount of care will guarantee a complete absence of errors, but a few procedures may limit their damage:

- Exert maximum care during the collection: always think that once a patient is gone so are the chances to record his/her data;
- Scan visually the data as they are collected to spot blatant problems;
- Use, if possible, an automated program to detect data that are out of the reasonable range;
- Prepare simple graph that can easily highlight "outliers". These can then be analyzed to discover the reason for their unusual characteristics;
- Be prepared to discard data that cannot be corrected.

However, never discard data simply because they look strange: they may in fact contain the key to your problem. Discard them only if you have reason to believe they are contaminated or invalid. Similarly, do not "fix" data once you discover the problem: it is often impossible to determine how and how much the error has changed the data. By fixing them one ends up analyzing one's own data rather than those collected. Unusual observations are often called "outliers" (Fig. 4 - 6). Formal methods have been proposed to detect outliers (Anscombe 1960; Barnett and Lewis 1978), but there is still much controversy surrounding them and little agreement on how to define an outlier.

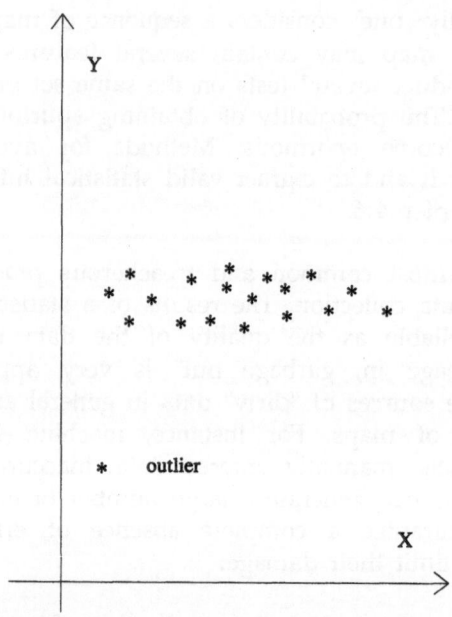

4-6:   Example of outlier. The isolated point may be an unusual member of the population or could be an error. Its presence is bound to strongly affect calculations of correlation, regression, mean etc.

4.3   NORMATIVE DATA

The first step towards understanding what is abnormal is to study what is normal. If we believe that the presence of a certain map feature is indicative of a certain disorder, what do we know about its frequency of occurrence in the healthy population? If the magnitude of a certain feature seems excessive, how high can it get in the healthy population? Answering these questions leads to the construction of "normative data", information about the behaviour of a variable in the normal population. Notice that "normal" may mean something other than healthy, as it happens when one tries to distinguish, say, a known form of epilepsy from a different one under investigation.

Constructing normative data is equivalent to estimating the "parameters" (Larsen and Marx, sec 5.2) of a population, parameters being any relevant quantitative features of that population. In general two types of parameters are needed to provide an informative description of the data at hand: "location" parameters, such as the mean and the median, which indicate the "average" magnitude of the features, and "dispersion" parameters, such as standard deviation and variance, which indicate how spread-out the data are. When one is analyzing single features of a map, then the usual measures can be used: mean, median and quartiles for location; variance, standard deviation, range and interquartile range for dispersion (ibid., Chapter 4). Advantages and disadvantages of each should be understood and analyzed before making an appropriate choice (ibid., Chapter 5).

When estimating multivariate features, like instantaneous maps or even maps over a sequence of time points (as in EPs), one can still use the same measures of location, but a problem arises for the variance structure. We mentioned earlier that for multivariate data covariances should also be evaluated, as they are used in most tests. However, the larger the number of variables under investigation, the larger the number of covariances to compute and hence the larger the number of observations needed for accurate estimation or even to obtain a non-degenerate matrix. (A variance-covariance matrix is "degenerate" if its determinant is 0. In that case the matrix cannot be used for statistical analyses). For instance with an array of N electrodes, one must estimate N variances and $N(N-1)/2$ covariances, for a total of $(N^2 + N)/2$ parameters. For $N = 2$ one has 2 variances, $2(2-1)/2 = 1$ covariance for a total of $2(2 + 1)/2 = 3$ parameters. If $N = 4$ one deals with a manageable 10 parameters, but for the 19 electrodes of the 10-20 system one needs 190 parameters. This means that 190 observations are needed simply to get a non-degenerate estimate of the variance-covariance matrix. And we have not even considered temporal correlations!

It is clear that for this type of problems a strictly accurate use of multivariate statistical techniques is not realistically feasible. And yet an estimate of the variability must be undertaken, since location parameters by themselves do not provide sufficient information on the range of "normality".

By simply ignoring correlations and treating each channel and each time point within an epoch as an independent entity, average maps and maps of dispersion parameters (usually standard deviations) may be constructed (as an example see Fig. 1 - 41 and

Fig. 1 - 42). The advantage of this approach is the simplicity of calculation and storage of the data and the possibility of displaying these maps on standard mapping equipment. For this reason this approach has become popular and adopted by several manufacturers of mapping systems. The disadvantage is the loss of the important information related to covariances and the corresponding inaccuracies that occur when performing statistical tests (see later paragraph). This approach is the only one feasible when analyzing physiological phenomena that require a large electrode array and an extended temporal observation. For instance VEPs recording the spread of a signal from the frontal to the occipital region for the detection of abnormalities in the conduction pathways are of this type.

For certain specific problems involving smaller areas, however, it may be possible to select a few channels and restrict the analysis to them by computing covariances (and/or correlations) as well. With this approach, however there is no natural method of display, although certain correlations may be selected for special maps. For instance if only the posterior head region and correlations between homologous channels are of interest, then such correlations may be displayed on lateral views of the scalp (Fig. 1 - 46b).

Normative data can be used in two ways. The first is to compare them to data collected from a different group in order to assess the presence of statistical differences. This will be discussed in the next paragraph. The second is to compare it to individual observations from specific patients and will be discussed now.

If no correlations have been considered, the comparison is done on each channel and each time point by computing how many standard deviations (sd) the observed voltage (v) is away from the corresponding mean voltage (u), that is, by computing $(u - v)/sd$. Assuming a normal distribution for the normative data (or a normal approximation) a threshold value of two standard deviations is customarily used to classify an observation as abnormal (point 2 in Fig. 4 - 7). This is because in a normal distribution 95% of the observations are within two standard deviations of the mean u. For example in Fig. 4 - 7 point 1 is within 2 standard deviations, while point 2 is just outside this range, being greater than the mean. However it must be remembered that this rule is not valid for non-normal distributions, such as skewed distributions, and that this rule still classifies 5% of normal subjects as abnormal. In brain mapping

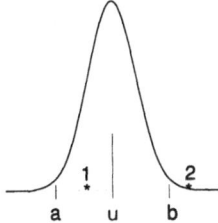

4-7:   From the normal distribution drawn, point a denotes the lower 5% bound, point b the upper 5%, u the mean. Observation 1 would then be classified as normal (within 2 sd of the mean u), while observation 2 would be considered abnormal.

this approach is also known as "significance probability mapping" or "SPM" (Duffy 1981). The major drawback of this is the fact that by ignoring correlations one may tend to overemphasize the presence of differences that extend over space or time, when in fact such extension may be due to correlations more than to extended abnormalities. In other words the test is prone to false positive detection.   Moreover in practice this method seems to be insensitive to low voltage abnormalities, as may be the case of a dead channel.

An alternative approach to standardized distances can be obtained by using the non-parametric method of percentiles.   For each time point and channel one computes the upper and lower percentiles of interest, say the 5th and 95th (Fig. 4 - 8) and classifies observations outside of these two limits as abnormal. This method, used often in the context of growth charts, works well in the presence of skewed distributions, but it still ignores the confounding effect of correlations and is more complex from the computational point of view, so that its computerized implementation presents more problems. A generalization of this idea is to estimate completely the distribution of a variable. This would allow one to precisely locate one observation at a given percentile and thus get a more precise assessment of its "abnormality". When the distribution is normal this reduces to estimating mean and standard deviation (Larsen and Marx Chapter 7), but for skewed distributions more complex non-parametric methods are needed (Lehmann 1975 sec 4.6). The problem becomes even more involved for the multivariate case and requires large samples. There are methods aimed at estimating percentiles with good accuracy from a small sample (Efron 1982) but these are rather

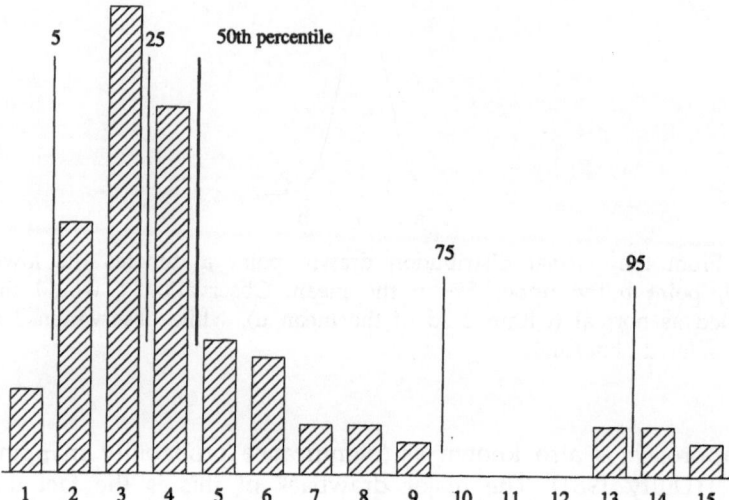

4-8:   An example of skewed distribution with a consistent group of high values that cannot be dismissed as outliers. There is no indication of the clump of data at the far right. On the other hand the median (x = 3) and the 5, 25, 50, 75 and 95% lines give a more accurate description of the characteristics of the distribution. This type of data is more common than one may think, due to mixing of factors in real-life observations.

involved and should be used only in special circumstances under the guidance of a statistician.

If correlations have been computed, one should use the "Mahalanobis distance" (Morrison 1976, Chapter 6). This is simply a generalization of the standardized distance which uses mean vectors and variance-covariance matrix instead of mean value and standard deviation.   Realistically this method can accommodate one time point at a time by viewing all channels simultaneously as the observation vector.   This still leaves the time correlations unchecked and does not provide information about the location of the areas most responsible for the abnormalities.

In a later chapter we shall see that by approaching this problem as one of classification many of the difficulties presented so far can be circumvented.

One problem associated with normative data is how to present them in graphical form. Univariate and bivariate histograms are quite effective and easily understood and can be drawn by several computer programs. However, as for the problem of estimating the entire distribution, a large sample is needed to draw an informative histogram. Moreover histograms for multivariate analyses of more than two variables are not possible. Some valid alternatives to histograms have been developed in recent years to represent data graphically. One of them consists of representing the distribution of a population by using some special percentiles. If one marks on a scaled line the 5th, 25th, 50th, 75th and 95th percentile, the result is a five-point diagram showing the central location of the distribution as well as its degree of variability and skewness. It has become customary to beautify this diagram by constructing a rectangle going from the 25th to the 75th percentile and adding whisker extending to the 5th and 75th. This is called a "box-plot" or "box-graph" (Cleveland, sec 3.2 and Fig. 4 - 9) and can be a very valuable graphical tool both for exploratory work and for reporting data summaries. Several variations of the box-plot have been developed to serve specific illustration purposes. These variations can be easily read and interpreted case-by-case.

To represent binary relations a "scatterplot" is the standard tool (the graphs used in Fig. 4 - 4 are examples of scatterplots). Its generalization to multivariate analysis is called a "scatterplot matrices" (Cleveland, sec 2.5) and both these methods can be quite effective. They can be effective in highlighting interesting relations and patterns of relations. As a simple example, to study the relationships among the response latencies to a VEP of O1, Oz and O2, one could prepare a scatterplot matrix and visually assess it.

One final type of estimation relevant in the construction of normative data is "regression" (Myers 1986). This is used when it is believed that one variable of interest is linked to another variable by some kind of functional relation. If this second variable is easier to measure, then knowledge of such relation can generate important information. This idea is behind most of the diagnostic procedures that use laboratory tests on one blood chemical characteristic, say, to obtain assessment of the clinical status of a disease.

The simplest type of functional relation between two quantities is the "linear" relation, where the two quantities, say x and y, are linked

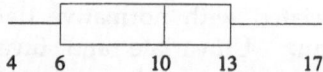

4   6   10   13   17

4-9:   A box-plot. The middle line corresponds to the median (10), the two outer edges to the 25% and 75% percentiles (6 and 13) and the whiskers extend to the 5% and 95% percentiles (4 and 17). Notice how the asymmetry is clearly visible.

by an equation of the form $y = ax + b$. If the two variables are linearly related, then "linear regression" can be used to estimate the constants a and b in the equation and thus obtain a workable relation between the variables (Fig. 4 - 10).

Because of the simplicity of the calculations needed in linear regression, this method has become quite popular, but, once again, certain conditions have to be met before using it. The most important one is that to use linear regression one must assume beforehand not only that there is a relationship between the variables, but that this is linear. If this is not true than the results can be quite misleading (Fig. 4 - 11). Moreover the usual requirements of normality, independence and equal variance for the errors must be satisfied. Finally, the relation is usually reliable only in that range of x values for which data have been collected, while extrapolations outside of these values are not warranted.

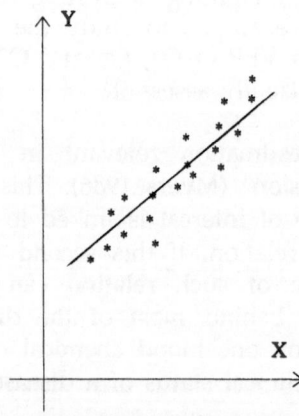

4-10:   X and Y seem well linearly related. The line showed is the regression line estimating the true linear relation between the two variables.

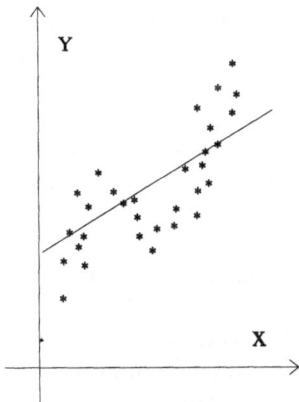

4-11:   Here a regression line is found, but the relation between X and Y is clearly non-normal and has a definite non-linear form, so the value of the regression line is debatable.

As one might expect, methods exist to estimate functional relations that are not linear as well as for estimating relationships among more than two variables ("multiple regression", see Draper and Smith 1981). The difficulties present in simple linear regression increase considerably when dealing with multiple regression and new problems and complications arise that must be dealt with. For a full discussion of such problems the monographs by Morrison (1976) and Draper and Smith (1981) are excellent sources.

4.4 STATISTICAL COMPARISONS

In order to compare   two or more clinically different groups (e.g., normative data and a clinical group) one needs a statistical test able to separate the natural variations in the groups from the systematic differences among them. For instance, if a spike with a certain peak topography is observed in two clinically different groups of patients, we may want to know whether the mean amplitudes are equal in the two groups. Or we may want to know whether the mean latency of the response to a new stimulus is equal to that of a standard type of stimulus. By comparing two topographies associated with two given phenomena or to the same phenomenon in two different groups of subjects one is posing the same problem in a multivariate setting.

Although many tests exist that answer such questions, they all follow the same general pattern. The observed data are used to construct a test value which, if equality is present, should have a known distribution. The likelihood of the test value under this "null hypotheses" of equality (i.e., the "p-value") is then compared to some pre-set threshold probability and if the computed probability is lower than such value, the observations are judged to be too unlikely to happen under the assumption of equality which is therefore rejected. Since in this way one is testing the hypothesis of equality, these techniques are known as "hypothesis tests".

It is important to remember that the statistical tests we shall discuss here only deal with the question of whether the data support the existence of a difference, ANY difference. This means that sometimes such difference, if it exists, may be practically irrelevant, in that it may not generate a good separation between the two groups in question. Another point to remember is that the threshold probability against which one compares the observed one, is arbitrarily chosen. The values of 0.05 and 0.01 are often used, but they became popular only because prior to computers, when statisticians used to rely on tables, these values seemed reasonable standards to be used. In practice one should be careful not to give these two values any sacred meaning and be prepared to assess the "statistical significance" of an experiment on the basis of other factors as well.

In the univariate case, if the distribution of the variable is normal, the most common such test is an appropriate "t-test" (Larsen and Marx, Chapter 8). A "one sample" t-test checks whether the mean of a population equals a set value. A "two sample" t-test checks whether the means of two different populations are equal. However, if in a two-sample situation the values are observed in correlated pairs one uses the "paired" t-test, which is just a one-sample test performed by comparing the paired differences to the value 0. This occurs, for instance, when analyzing the same variable before and after a treatment or a physiological event (e.g., EEG spectrum before and after a sedative).

All t-tests require a normal distribution, but work reasonably well also for moderate departure from this assumption. The two-sample t-test requires both populations to have similar standard deviation. This does not mean that the standard deviations of the two samples should be the same (in fact they will always be different), but large differences in such sample standard deviations should be investigated. Alternative tests exist for the case where the populations are known to have different standard deviations. However their statistical properties are not as well known and hence they should be used with care (Mood et al. 1974, Chapter 9). A large sample usually moderates this kind of problem.

If the distribution is known, but not normal, it is often possible to construct ad hoc tests that are appropriate for that distribution (ibid.). This often turns out to be a very technical method that improves only slightly upon the power of the simpler and more generic tests known as "non parametric" (Lehmann 1975). These are tests that look only at the relative magnitudes, or ranks, of the observed values, rather than their actual magnitude, and do not depend on the distribution of the variables. To give an idea of how these tests achieve their generality, let us suppose that we want to find whether a certain value is the median for the given variable. If this were true one would expect about half of the observations to be larger than such value and half smaller, so the question can be reduced to a coin tossing experiment, whose probability distribution ("binomial") is well known and can be used to assess the likelihood of the observed number. This is known as the "sign test" (Lehmann 1975, Chapter 1) and is a one sample test that can also be used in a paired two sample problem. Similar types of methods are behind other more complex non-parametric tests.

To non-parametrically compare two groups for a difference in the mean the most common test is the "Wilcoxon rank-sum test" (ibid.), which analyzes whether the global relative ranks of the observations in the two groups are randomly distributed or are skewed in favour of one group. The Wilcoxon "signed rank" test (ibid.) deals with a paired two-sample question more powerfully than the sign test by analyzing the relative ranks of the magnitudes of the paired differences. It is even possible to test the more general question of whether two groups have the same distribution, via the Smirnov test (ibid.).

When more than two groups must be compared generalizations of the above tests must be used. Generalizing the t-tests leads to "analysis of variance" or ANOVA (Montgomery 1984). This is another quite popular technique based on assumptions that are frequently ignored, such as normality, independence and equal variance among different groups. One further difficulty of ANOVA is the fact that in its simplest and most used form it only tests for equality of the means among the different groups under comparison. When such equality is rejected by ANOVA one needs additional tests to verify where the differences occur, rather than a simple visual analysis of the means. Finally, in analogy with the two-sample vs paired-sample difference for the t-test, when several groups are compared the connection between them and the sampling method become critical in correctly setting up the statistical analysis and simple ANOVA is not always appropriate. A description of the different methods and techniques may be found in any discussion of experimental design, such as Montgomery (1984).

Analogs of ANOVA exist if one wants to analyze multivariate data non-parametrically ("Kruskal-Wallis" test) and for many other situations (Lehmann 1975).

In many instances the problem of an unknown distribution can be resolved by grouping the data into cells, each cell characterized by a range of values or even a qualitative feature. This generates a "contingency table" (Larsen and Marx 1981, sec 9.5) and the corresponding information can be analyzed in several ways according to the question of interest. The most common test associated with a contingency table is the "chi-square" test (Larsen and Marx 1981, Chapter 9). Although this test is often used as an all-purpose tool, it in fact asks a very specific question, namely whether the distribution of the data in the different cells is independent of the two groups under analysis. The detection of trends, or the determination of the "better" group (in whatever sense) should be addressed by other more sophisticated tests.

One important question connected with hypothesis testing is that of whether one should use a "one-sided" or a "two-sided" test (Ibid. Chapter 6). Most computer program provide p-values for both situations and it is up to the researcher to choose the correct one. The answer depends on the question we are asking. If we want to know whether a difference exists, then a two-sided test should be used.   If we want to detect a difference in a specified direction than a one-sided test is appropriate.

If one is dealing with a multivariate setting, the first instinct would be to simply apply a univariate test to each variable and then arrive at conclusions for each of them. Unfortunately the multiple comparison problem mentioned earlier applies here more than ever. If the null hypothesis is correct and we choose 0.05 as the probability level at which we reject the null hypothesis, an average of 5% of the tests performed will give significant p-values by chance only, that is, false positives. So the greater the number of comparisons, the greater the chance of false positives. Several methods exist to approach this problem correctly.

The simplest, but least effective, is to change the threshold value used on individual tests so that the overall probability of false positives is the one required. This is the "Bonferroni" method (Bickel and Doksum, sec 7.4) and is not very effective because of its low "power", or ability to detect existing differences. For this reason the Bonferroni approach should not be used when many variables are studied simultaneously. If the distribution of the variables under question is approximately normal, a more effective test is Hotelling's T-squared (Morrison 1976, sec 4.2). This test is a multivariate generalization of the t-test and uses the variance-covariance matrix to give rise to a single p-value, thus avoiding the multiple comparison test. For example, if one wants to compare the peak topography for certain spikes appearing in two different groups of patients, the T-squared test applied to the entire array of electrodes used would be the most appropriate. The drawback of this test is that it gives an overall significance probability, without specifying in which region, if any, the difference is most relevant. Post-hoc tests exist to answer such more specific questions and each of them is designed to address a specific question related to individual variables or combinations of them (Montgomery 1984, Chapter 3).

In some instances the experimenter knows that the variables under study are very strictly related and in fact they all depend on a much smaller set of variables. In this case it may be useful to extract such smaller set in order to reduce the problems related to high dimensionality and also to better understand the nature of the variables recorded. For instance for certain spontaneous discharges the potentials recorded at an array of 19 electrodes may all depend on one or two focal points and be explained in terms of them. This is an example of the so-called "inverse solution" problem in EEG. One approach to this problem is to formulate a specific mathematical model explaining the way in which the potentials are generated and then use data to estimate the parameters of this model (see Part 2). If no model seems satisfactory two classical statistical methods are available.

The first method is that of "principal components" (Morrison 1976, Chapter 8). It consists of finding that linear combination of the observed variables that accounts for most of the variability in the data. This combination defines, in the multidimensional space of the variables, a direction called "first principal component" that sometimes may have a practical interpretation (Fig. 4 - 12). For instance in the situation described previously, if the first principal component involves mostly the occipital electrodes it may be interpreted as a focus in the occipital region. The next step consists of finding another linear combination, perpendicular to the first principal component, that accounts for most of the remaining variability. This is the "second principal component". They may also have an interpretation as a second focus or a secondary effect of the first focus. The process can be continued by finding as many principal components as there are variables. If the method is successful, however, the first 2 or 3 components may well account for a large percentage of the total variability and there is no need to proceed further. The principal components that are deemed of interest can then be used in subsequent studies as the variables of interest, thus reducing the dimensionality of the problem.

There is a certain degree of arbitrariness in the choice of how many variables to retain and this may affect the validity of the study. Some computer packages that perform principal component analysis contain automated algorithms for the choice of the components. One must remember, however, that such algorithms always contain, at some point, an arbitrary element of decision and it may be important to know, and to report, what that is.

The second method is that of "factor analysis" (Morrison 1976, Chapter 9). Here one fixes a priori the number of independent factors believed to have generated the variables recorded and then uses the data to find the linear combination that best explains the data in terms of the desired number of factors. For instance if an array of 10

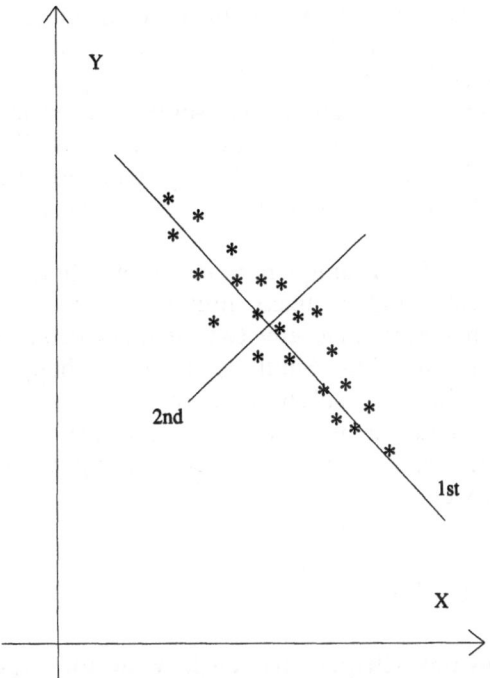

4-12: The longer line is the first principal component, the shorter one is the second in this two dimensional example. Each point can be represented by its projected positions on the two lines. If the first line accounts for most of the variability, as in this case, the dataset may be reduced to a one dimensional dataset.

electrodes is recording a phenomenon believed to be generated by only two foci, one would construct the 10 x 2 matrix that best explains the data in terms of the two foci. There are several theoretical problems associated with factor analysis and for this reason factor analysis is often a source of controversy and suspicion. The two main problems are the choice of the number of explanatory factors and the non-uniqueness of the solution.

With respect to the first, we should notice that the number of factors is chosen arbitrarily. Factor analysis of the same data based on different numbers of factors may lead to rather different, if not contradictory results. This can be puzzling when the method is used to discover mechanisms of action or to confirm certain hypotheses. Once again, some statistical packages boast automatic selection of the number of factors, but this has to be done using an arbitrary criterion which, although based on some sound rationale, may not be appropriate for the question at hand.   Moreover, such automatic selections are data based, so that they tend to provide the most optimistic, but not always the most accurate choice.

The problem of non-uniqueness of the solution means that for a given set of data and a fixed number of factors several "best" transformation matrices exist, any two of them being related by some high dimensional rotation. Similarly to what happens in the first problem, these different solutions may be interpreted in different ways and no standard method of choice exists (Morrison, 1976, Chapter 9). The previous warning about computer packages claims applies here as well.

4.5 CLASSIFICATION

In the previous Chapter we dealt with the types of statistical problems that appear most often in the literature. In medicine, however, the problem of classification is at least equally important and one could argue that it is much more relevant. A fundamental part of a physician's knowledge is used to identify the disease affecting a given patient by using quantitative and qualitative information about that patient's physiological state. Methods aimed at statistical comparison are not appropriate to solve the classification problem, even though they are often used for that purpose. To clarify the difference recall that in statistical comparison one may ask whether the average value of a variable is different in two groups of subjects.   However if they are, the difference could be so small with respect to the existing variability that a great degree of overlap may exist between the two groups and that variable may not be used effectively to separate the two groups (Fig. 4 - 13). On the other hand the average values may be equal and yet one group have more variability than the other, so that an extreme value could be easily associated to the more variable group (Fig. 4 - 14).

In classification we look for those variables or combinations of variables that are best able to identify which group a given obser-

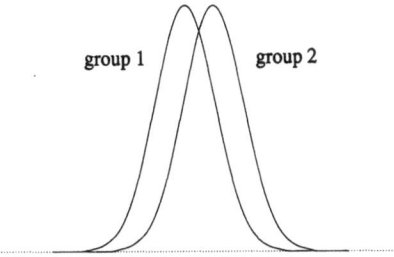

group 1   group 2

4-13: The two populations (groups 1 and 2) are different, but have a great degree of overlapping, so that it is difficult for any classification rule to distinguish between them. Also, the difference may be insignificant from the practical point of view.

vation belongs to. In a medical setting, when the classification is made among different diseases and leads to different diagnoses and therapies, it is also important to obtain, from the classification rule that one constructs, an indication of what physiological processes are explained by the rule and in what way the variables used in the rule are associated with the different conditions.

In the classification problem we have two or more groups, or "classes" and a number of observed measurements, or "explanatory variables". The problem is to set up a rule that, on the basis of the explanatory variables, will allow us to assign each observation to one of the classes. For instance, we may want to use some variables obtained from an EEG to decide whether a given patient is normal, has meningitis, has a tumour or a blood clot. What one wants to achieve is a rule that makes few mistakes, or that has a low error rate. If one of the groups to be classified is particularly important (e.g., healthy subjects), it may be interesting to look at the rate of correct classification of observations in the group ("sensitivity") and outside the group ("specificity"). It is important to understand that in some instances a certain degree of overlap among the groups will remain for whatever rule we choose and that the most we can hope for is to find an optimal rule, on the basis of the variables observed. In the simple case of two groups and of explanatory variables having a normal distribution, a standard method of obtaining a classification rule is by using Fisher's "linear discriminant function" (Fig. 4 - 15). This is simply that linear combination of the variables recorded that produces the best separation between the two groups (Morrison 1976,

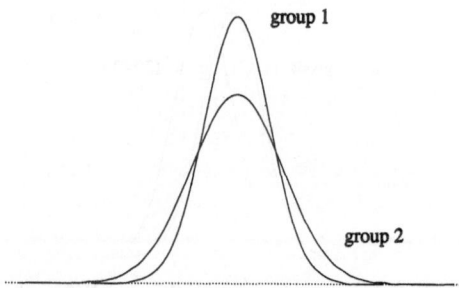

4-14: These two populations have the same mean, but different variability, so methods of classification based on equal variance would be useless (e.g. t-test), even though extreme observations could be safely classified in the more variable group.

Chapter 6). In fact if the hypothesis of normality is true, it is possible to prove that this is the best rule, in that it provides the highest sensitivity and specificity obtainable in the given situation. This property has made linear discrimination a favourite method and it can be found in many computer packages. When the situation allows it, it should be the method of choice.

There are, however, some drawbacks to this method. First of all the method ceases to be optimal when the distributions are not normal. So if some of the variables have a known skewness or, even more blatantly, if some variables are nominal (such as types of discharges), it cannot be used effectively. Moreover, even if the distributions are normal, the rule depends heavily on the variables used, in the sense that the introduction of a new and perhaps irrelevant variable may alter the formula for the rule and, with it, the interpretation. This is not desirable when one wants to extract from the formula some information about the relation among the variables used and the conditions under study. Another drawback is the fact that the resulting formula is usually technical in nature. For instance, if S denotes serum glucose, A denotes age in years and F is a variable equal to 1 if there is fever, or to 0 if there is no fever, linear discrimination may be based on an aggregate score formula that looks something like:

$$d = 0.342S + 1.265A - 0.023F \ldots$$

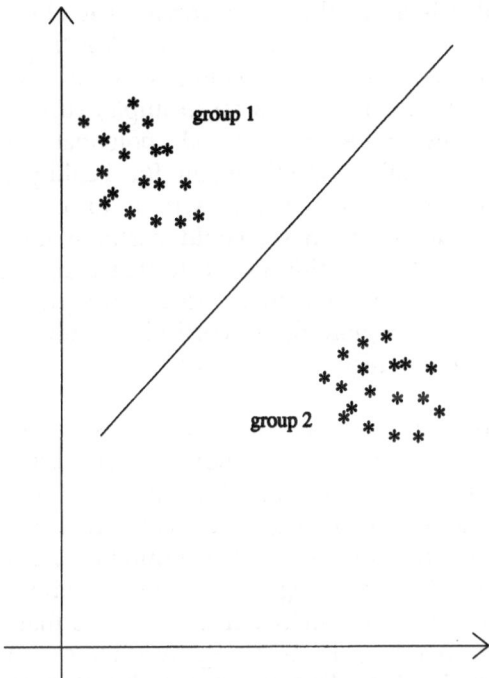

4-15:   Linear discrimination assumes that the two populations (groups 1 and 2) have different means but equal variance-covariance matrix. The classification is obtained by finding the line that best divides the two groups.

and as such cannot be easily computed in a clinical setting, nor does it follow the standard format of the diagnostic process.

Physicians are more used to rules of diagnosis that look like: "if serum glucose < 300 and age > 25 and there is fever then ...", which still can be written mathematically:

if S < 300 , A > 25 and F = 1, then ...

but have a much clearer interpretation.

This approach, however, does have a mathematical formulation, by means of "partitions" (Breiman et al. 1984, Chapter 1). A partition of the set of all values for the given variables is obtained by dividing such set into smaller distinct subsets. A classification rule can be obtained from a partition by associating with each of the subsets one of the groups to be identified. As an example, suppose that a certain visual stimulus generates an evoked potential response that is maximal in the central occipital region for healthy subjects, in one lateral occipital region for patients with a tumour and frontal for cortically blind patients. Then we could partition the scalp as in Fig. 4 - 16 and consider the classification rule that assigns an observed EP to the healthy group if its maximum occurs in region I, diagnose it as a tumour if it is in region II, as cortically blind in region III and uncertain in any other region.

But how does one derive the boundaries for such a partition? Once again, if the variables have a normal distribution in each group, then an optimal method has been known for a long time and it is based on the Mahalanobis distance mentioned earlier. With this method the partition is obtained by determining the mean position in each group and then drawing boundaries around them so points within each boundary have Mahalanobis distance that is minimum for the group corresponding to that region (Fig. 4 - 17). But normal distributions are hard to find as shown by the previous example, where the position of the maximum can only be one of the electrodes, so it is not even continuous. In that case optimality properties are no longer valid and other methods may provide lower error rates. Moreover the Mahalanobis distance method still requires the calculation of distance scores that often do not have a practical interpretation.

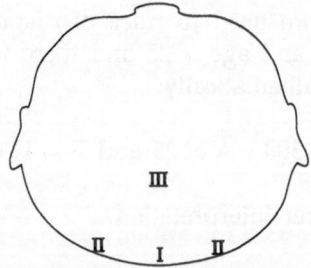

4-16: Regional discriminants based on scalp locations for the example of VEP data.

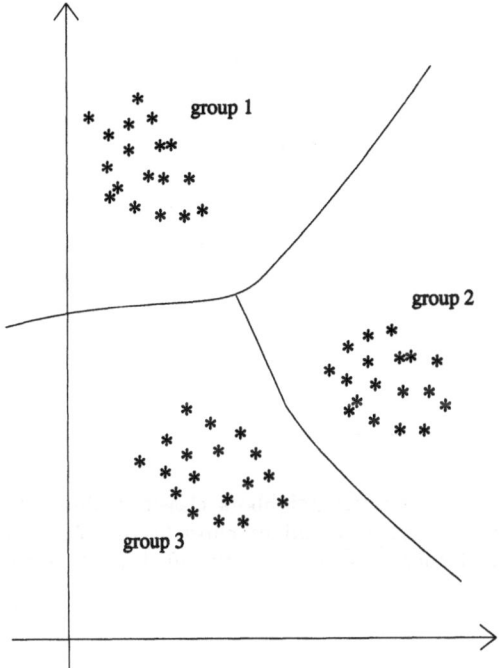

4-17:   When more than two groups must be classified, using Mahalanobis distance divides the region into several subregions, each being most likely to contain observations of one group.

Many alternative methods have been developed, but they are only starting to become popular now, mostly because they require large amounts of calculations that are feasible only by means of computers.   We shall mention only two of them, the "k-nearest neighbours" and "classification trees".

In these method, as in all others, the rule is constructed on the basis of a sample of observations of which we know the classification. For the method of k-nearest neighbours, in its simplest form, the values of the variables are used as coordinates in a high-dimensional space, and an arbitrary integer k is chosen. Any observation is then classified as belonging to that group which contains the majority of

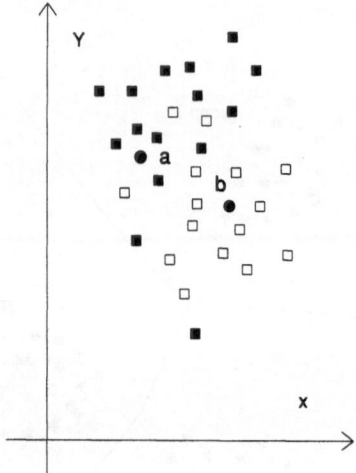

4-18:   The method of k-nearest neighbors classifies observations according to which group has more representatives close to the observation in question. For instance if k=3 then x would be classified as a filled square, y as a hollow square.

the k sample points closest to the observation in question (Fig. 4 - 18).   A complex analytical procedure can be used to determine the partition associated with this classification. Although this method has a good theoretical basis, related to the idea of "density estimation" (Silberman 1986), it has some practical problems that must be controlled. First of all k is arbitrary and its value may change the final result; then the partition rule changes if the coordinate system is changed by, for instance, changing the units of measurement of one variable. In an EEG setting, one may get different partitions when trying to construct a classification rule based on potentials measured in mV or uV. Again the resulting rule may not be easily interpreted.

The method of "classification trees" is perhaps the most general and appropriate for obtaining rules that are independent of units and distribution and are easy to interpret. A classification tree consists of a branching system of decisions, usually stated as questions about the variables measured (Breiman et al. 1984, Chapter 2). Each observation is sifted through the system and sent to a "terminal node" according to the answers it provides to the given questions. Each terminal node is associated with a group, so that the observation is assigned to the group of its destination node (Fig. 4 - 19).

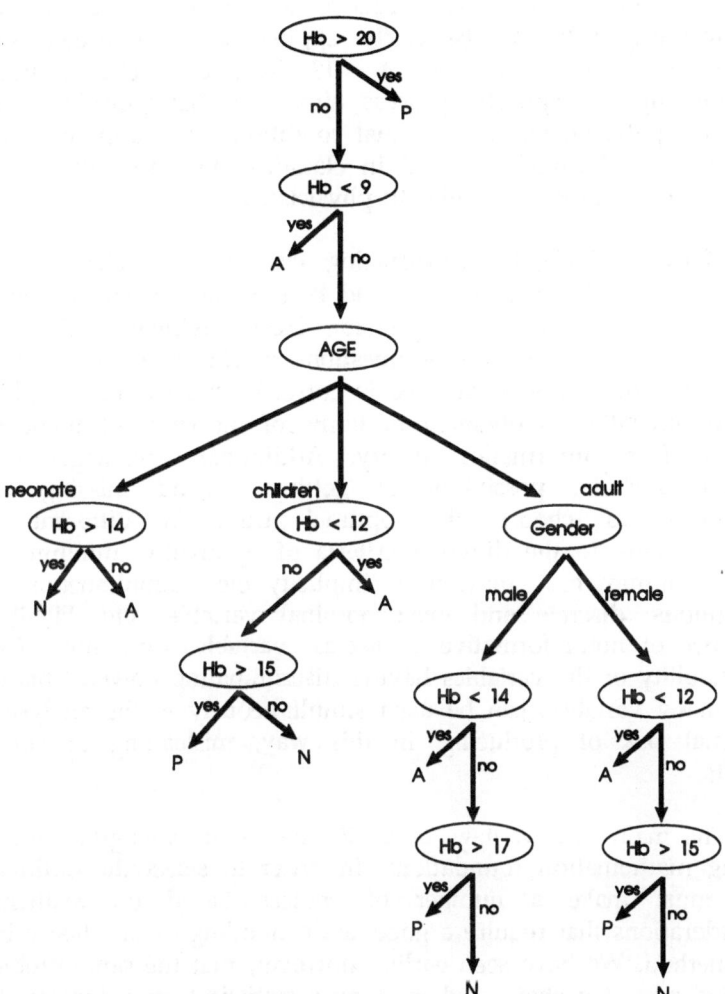

4-19:   This classification tree provides a way of describing the diagnostic process followed to decide between anemia, polycytemia and normality on the basis of hemoglobin level (Hb), age and gender. N = normal, A = anemia, P = polycythemia.

Because each question can be phrased in terms of a single variable or a simple combination of them, each terminal node can be often described in qualitative terms related to those questions, so that the resulting rule can be easily applied and interpreted by the practitioner. For example Fig. 4 - 19 describes a classification tree representing a diagnostic process aimed at distinguishing between anaemia, polycythemia and normal conditions. It is apparent that the mathematical formulation used in classification trees follows closely the diagnostic process familiar to physicians.

Many methods for constructing classification rules have been proposed, but the most general and best mathematically grounded is CART (Classification And Regression Trees, Breiman et al. 1984). In this method the sequence of questions is chosen by optimizing, at each step, the ability of the tree to correctly classify the sample data. The optimization is obtained in terms of measures of homogeneity coming from information theory. Additional advantages of this method include: possibility of getting around missing values; possibility to choose the optimal trees in different ways, corresponding to the different criteria of optimality (minimum error rate, minimum cost, maximum simplicity etc.); simultaneous use of continuous, discrete and even nominal variables etc. Finally, the presence of non-informative or useless variables does not affect the detectability of the variables having discriminating power. This means that many variables can be used simultaneously in the analysis with minimal risk of producing, in this way, misleading or confusing results.

The main disadvantage of CART lies in its strength, namely, its strong mathematical foundations. In order to select the optimal tree one must make a number of choices based on mathematical considerations that require a good understanding of the theory behind the method. We have seen earlier, however, that the same problem of making educated choices when using a statistical procedure applies to almost any analysis, so that the contribution of a competent statistician is essential when using CART as much as when using more popular procedures.

Another example of classification rule obtained by applying CART has been described (Wong et al. 1989). Several EEG features and variables were used to differentiate between two types of benign rolandic epilepsy of childhood. The classification rule constructed through CART uses three variable out of an original set of 10 and has an estimated 80% rate of correct classification.

4.6 EXPLORATORY VS. CONFIRMATORY ANALYSIS

Because the results of any statistical analysis are based on a sample, they are subject to variability and hence to errors. As a consequence, the eternal question accompanying every such analysis is whether the data analyzed allow conclusions one way or another. Is a p-value lower than .05 sufficient to claim "conclusive" evidence? Should we use .01? Was the sample size large enough? Was the statistical power of the experiment high enough? Was the analysis a "designed experiment" or an "observational study"?

While mathematics can be used to answer many such questions, philosophical differences of opinion still remain among theoretical as well as applied statisticians as to the interpretation of the results of a statistical study. Caution is mandatory if one is to remain objective and avoid technical criticisms.

The first question one should address is whether the data were collected and analyzed in response to a specific question or to a generic interest in a problem.

If a specific question had been addressed, an experiment designed accordingly, criteria of random selection of the sample respected and a theoretically suitable method of analysis used, then one can safely make conclusions from the results of the analysis, with the only proviso given by the ever present margin of statistical error. On the other hand, if the data were originally analyzed within a generic plan to find "interesting results" and one procedure gives rise to low p-values, or other exciting values, one should stay away from overly confident conclusions. In fact, under those conditions the same p-values one obtains do not have the same interpretation as probabilities that they have in designed experiments. The problems of random selection and multiple comparisons become very relevant in an observational study and the p-values become more "qualitative" and less "quantitative" indicators. This difficulty is particularly acute in brain topography, where the variety of problems to be studied and the lack, in many areas, of solidly established criteria for clinical assessment, lead researchers to use *large* amounts of data collected on *few* subjects to provide answers to *many* questions.

It is equally wrong, however, to dismiss such studies as irrelevant because of their observational nature. We may think of statistics as a collection of methods aimed at obtaining information out of the ignorance due to variability. In that light, the results of observational studies can provide much information at little cost and lead to important discoveries. As such they are worth performing and studying and should be viewed as methods for "hypothesis generation" rather "hypothesis testing".

Here are a few general criteria that should be followed when reporting or reading the results of statistical study.

1) The method of collection of data and the original goal of the study should be clearly stated.
2) The statistical method used should be well suited to the data and to the conclusions one wishes to draw.
3) Possible sources of confounding (see Chapter 4.2), of non-randomness in the sample and of multiple comparisons should be identified and discussed.
4) When estimates are provided, standard errors should always follow and when p-values are reported these should be provided in their exact value, not simply as being more or less than some pre-set value.
5) If the study is not clearly a designed experiment, a hypothesis suggested by the analysis should be stated and an experiment to confirm or refute such hypothesis should be described and proposed as the next step in the research.

A hypothetical example will serve to illustrate all these concepts practically. Let us suppose that asymmetry in the occipital response to a certain visual stimulus is believed to be associated with a given condition of the brain. After a statistical analysis of the available data, a report finds that a t-test applied to the correlation coefficient between left and right occipital channel responses provide a p-value less than .05 for the comparison between patients with that condition and normal subjects. Then one should ask:

1) Was the purpose of the study to analyze the effectiveness of the correlation coefficient in this clinical problem, or, more generally,

to search for some kind of laboratory test supporting the clinical diagnosis?

2)   Was the t-test well suited for the task, given its requirement of normality that is believable only if the correlations are not too close to 1 or -1?   Was this a hypothesis test or a search for a classification rule, in which case the t-test may not give usable information?

3)   Was the correlation coefficient the only variable under scrutiny or were there other measure that proved less reliable? This question would be essential to assess the validity of the conclusions. Also, how were the patient data selected? Was a patient entered on the basis of a perceived asymmetry or, rather, on unrelated criteria?

4)   Why was the threshold value of .05 chosen? How much lower than .05 was the observed p-value? What about other relevant measures, such as the correct classification rate of a rule based on the correlation coefficient? Given that a difference exists, did the report give an estimate of it and the precision of the estimate?

5)   If this finding was obtain in the midst of other results, what experiment should be performed to validate it? How many patients and normal subjects should be used? What test or statistical methods should be used?

Addressing this type of questions openly and clearly can provide much more information about a given clinical problem than large amount of data analyzed mechanically according to "popular" methods, rather than appropriate ones.

An excellent source of information on methods for exploratory data analysis is Tukey (1977).

PART 5

SELECTED NORMATIVE DATA

Selected examples of data collected from normal subjects are shown here for the purpose of illustration only, and not as a comprehensive atlas. The objective is to provide examples for beginners so as to facilitate their learning process. For this reason, only the following paradigms are included: flash VEP, pattern reversal VEP, and P300 auditory EP. The data is further divided into several age groups. This data have been provided by Bio-Logic Systems Corporation (Mundelein, Illinois).

All data were recorded using 20 channels (International 10-20 System positions plus Oz). The reference used was linked cheeks. The recording protocols were as follows:

1. Flash VEP

   Amplifier filters: 1 - 100 Hz
   Number of average: 200
   Stimulation rate: 1.0 Hz
   Flash source: strobe light
   Epoch: 512 msec.
   Age groups: 6-10, 11-15, 16-20, 21-30, 31-40 yrs.

2. Pattern reversal VEP

   Amplifier filters: 3 - 100 Hz
   Number of average: 200
   Stimulation rate: 1.7 Hz
   Pattern generator: 16" TV monitor
   Check size: 16H x 12V
   Distance: 1 m
   Epoch: 256 msec.
   Age groups: 16-20, 21-30, 31-40 yrs.

3.   P300 auditory EP

Amplifier filters: 1 - 100 Hz
Number of average: 200
Stimulation rate: 1.1 Hz
Common tone: 1000 Hz
Rare tone: 2000 Hz
Ratio: 4:1
Epoch: 512 msec.
Age groups: 16-20, 21-30, 31-40 yrs.

4.   EEG (resting eyes open and close)

Amplifier filters: 1 - 30 Hz
Individual epoch: 2 sec.
Epochs per FFT: 30-40 sec. artifact-free
Age groups: 11-15, 16-20, 21-30, 31-40 yrs.

The following points are of note:

Flash VEP

The main occipital positivity occurred at latencies between 132 - 146 msec., involving mainly Oz, less so O2 and O1, and dropping off rapidly anteriorly. It has a symmetric distribution. The later positivity at 200 msec. shown for the oldest age group has a more diffuse and anterior distribution.

Pattern reversal VEP

The P100 peak showed a sharp symmetrical occipital prominence, with little anterior activity. Source derivation gave a slightly more bilateral spread than flash VEP.

P300 AEP

All maps showed a broad symmetric peak centered at Cz. There is a tendency for more posterior involvement.

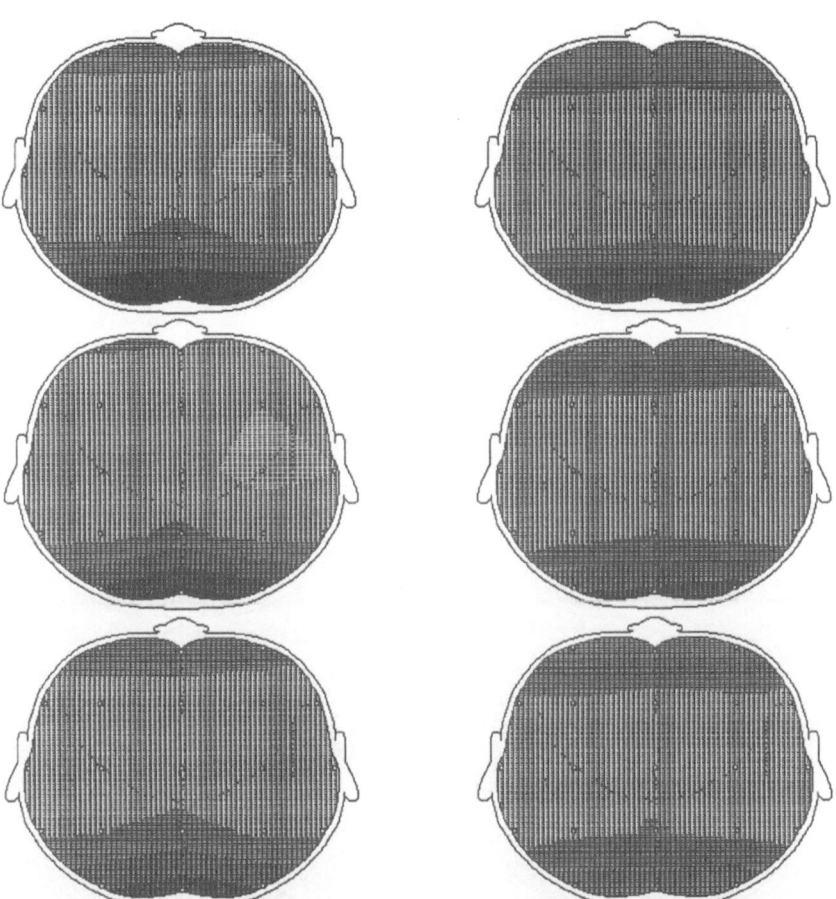

5-1a: Flash VEP: maps displayed at the time of peak occipital positivity. Left from top: 132ms (6-10 yrs), 136ms (11-15 yrs), 142ms (16-20 yrs). Right: same with source derivation. All figures in this section have black = positive, white = negative. There is some unavoidable stripling in the maps due to technical reproduction factors.

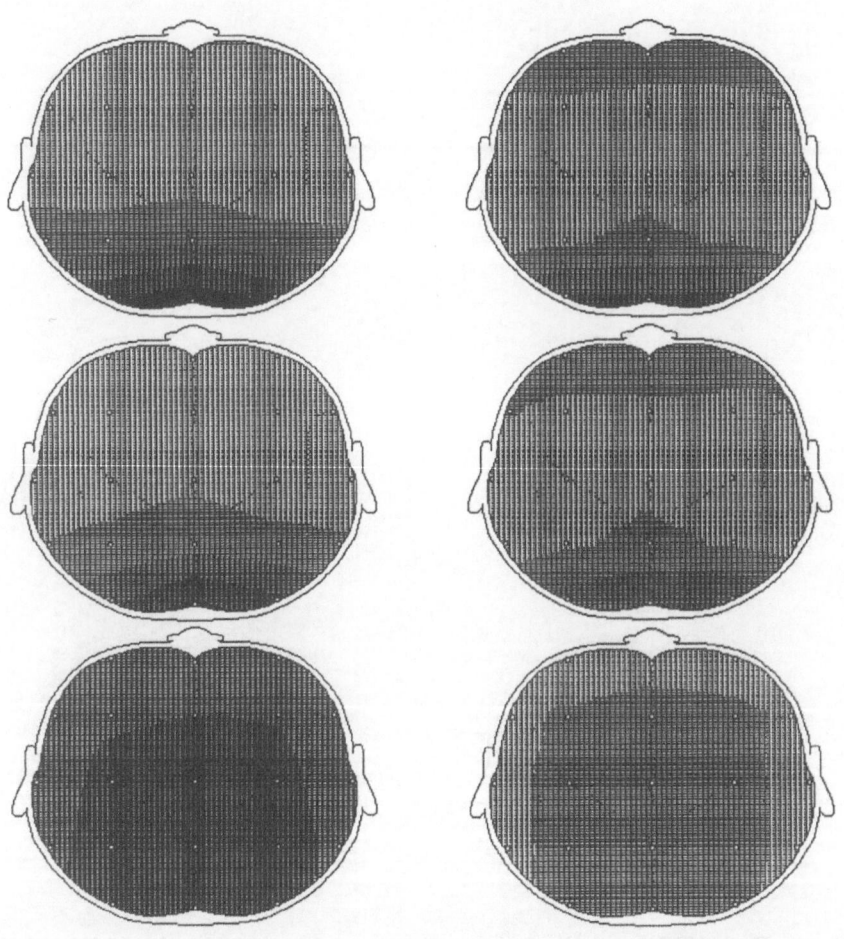

5-1b: Left from top: 138ms (21-30 yrs), 146ms (31-40 yrs), 200ms (31-40 yrs).
Right: source derivation.

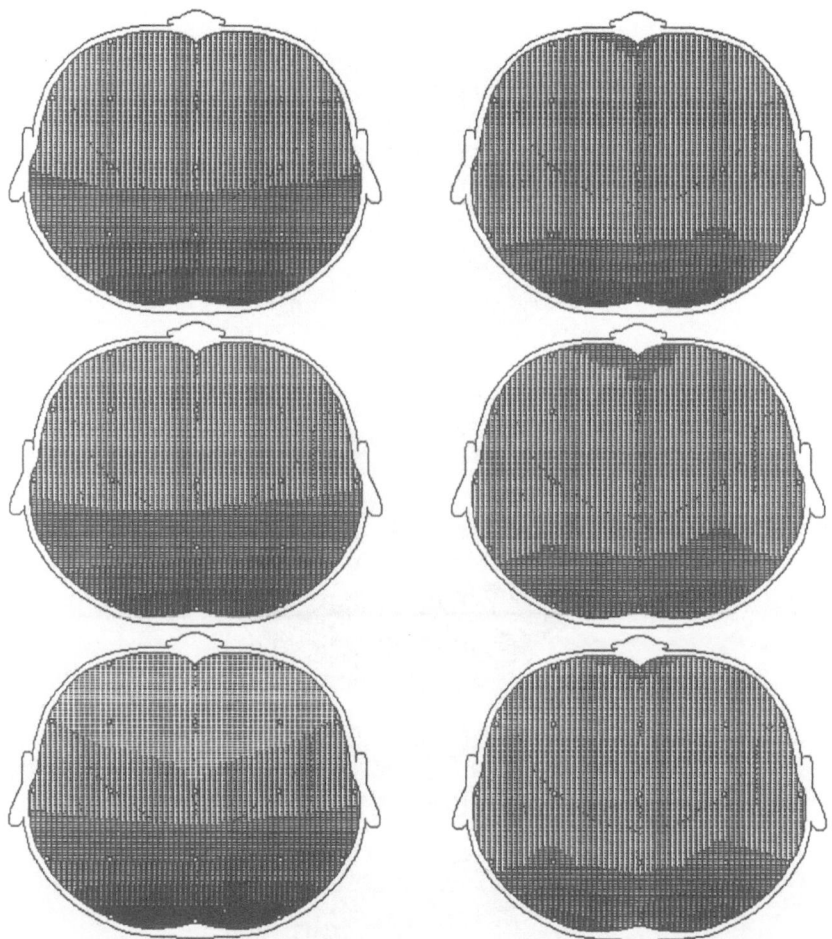

5-2: P300 peak. Top down: 308ms (16-20 yrs), 298ms (21-30 yrs), 304ms (31-40 yrs). Note the symmetric distribution.

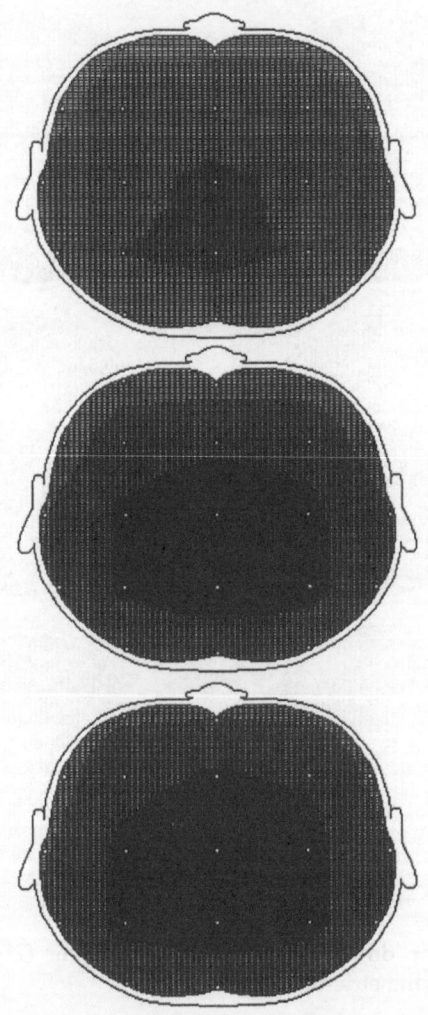

5-3:   Pattern VEP: P100 peak. Left from top: 101ms (16-20 yrs), 102ms (21-30 yrs), 102ms (31-40 yrs). Right: same with source derivation.

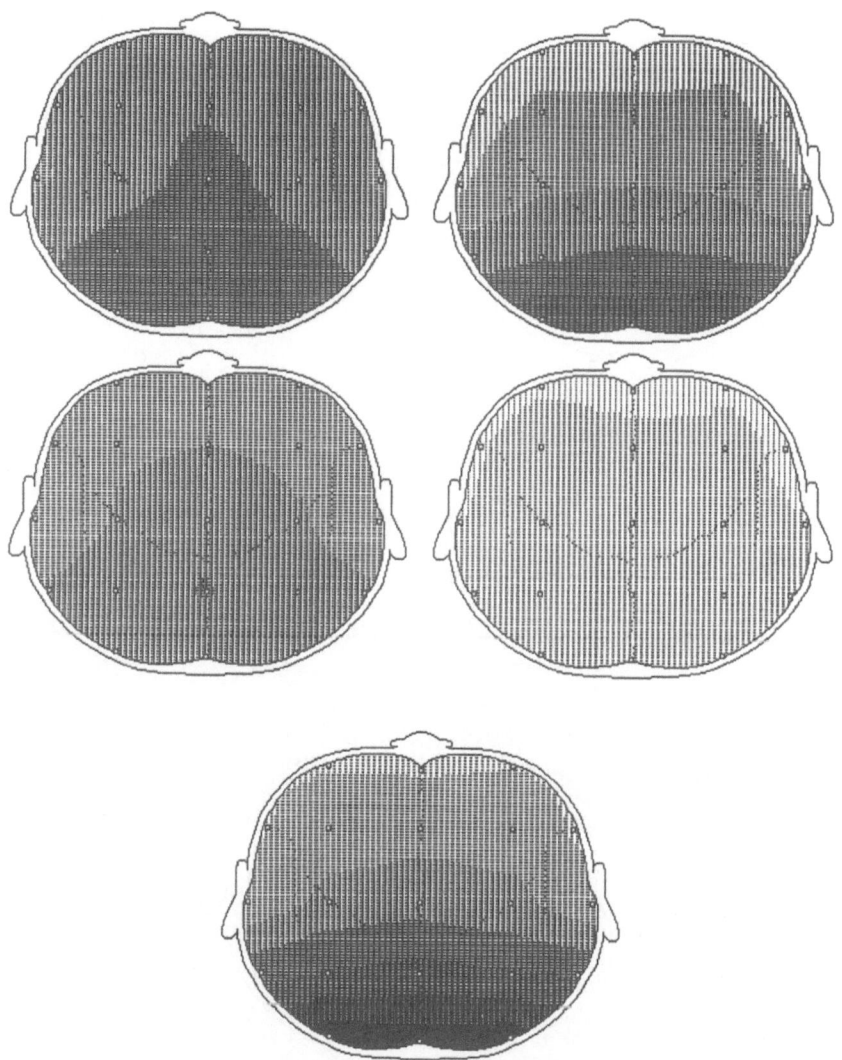

5-4a: FFT band maps (Eyes closed, 11-15 yrs). Top left: delta (0.5-3.5Hz); middle left: theta (4-7.5Hz); top right: alpha (8-12Hz); middle right: beta (14-18Hz); bottom: alpha peak frequency map. As FFT do not have negative values, black = high amplitude. Fig. 5-4a through fig. 5-4h all have the same display format and unipolar colour/voltage scale: black = high activity, white = low activity.

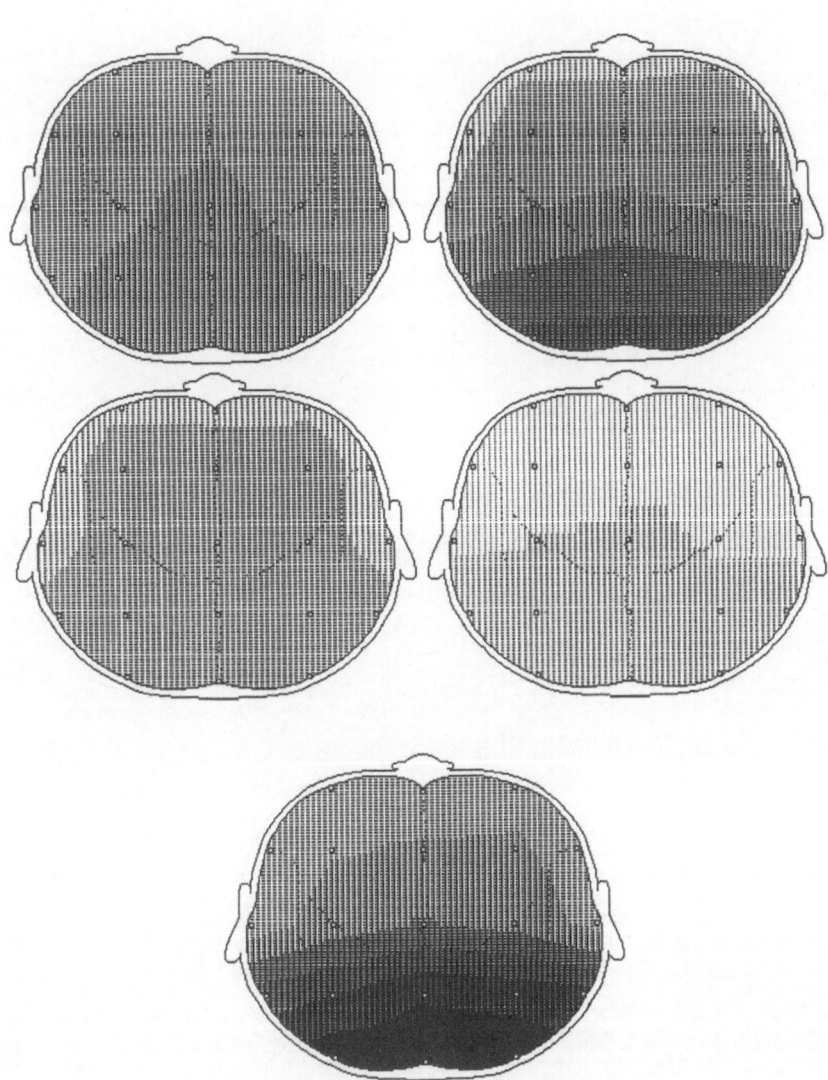

5-4b: FFT band maps (Eyes closed, 16-20 yrs).

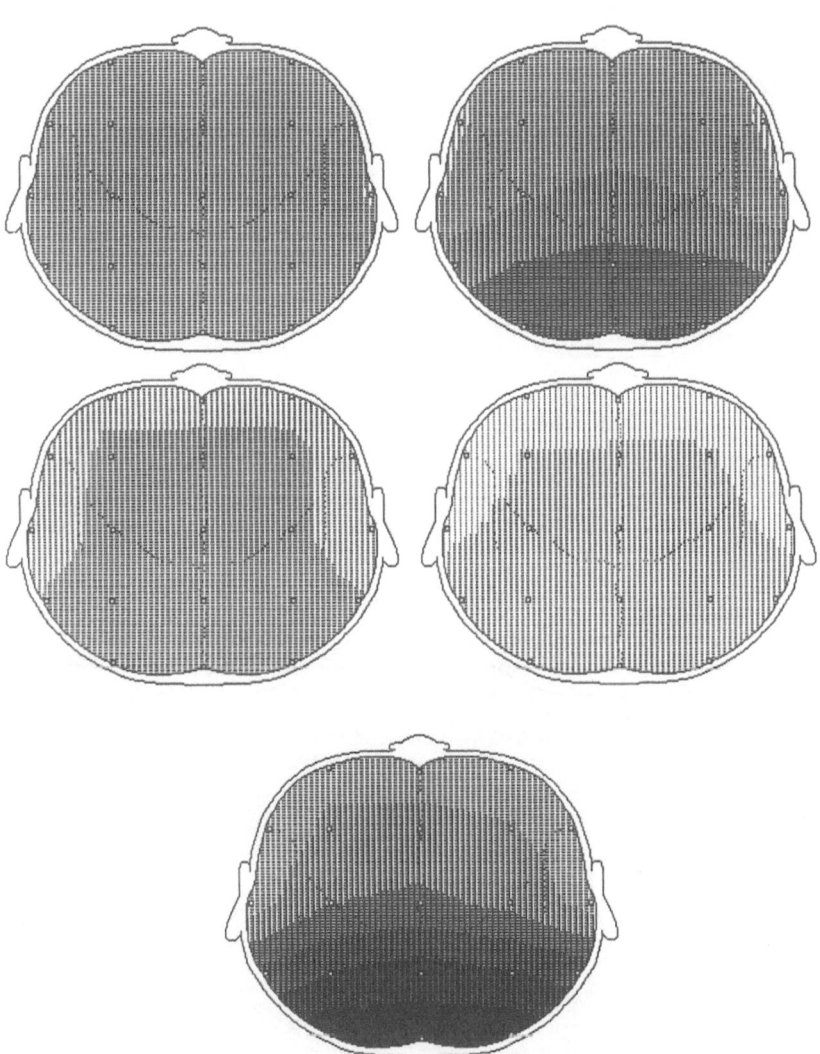

5-4c: FFT band maps (Eyes closed, 21-30 yrs).

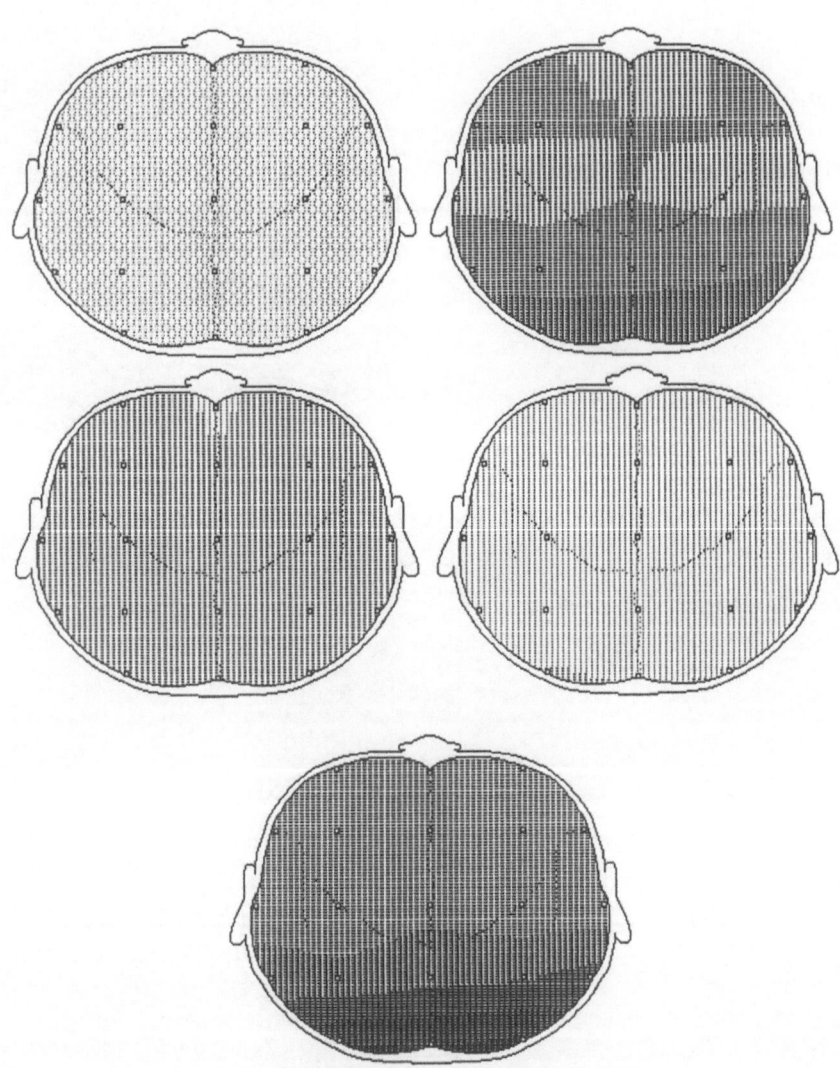

5-4d: FFT band maps (Eyes closed, 31-40 yrs).

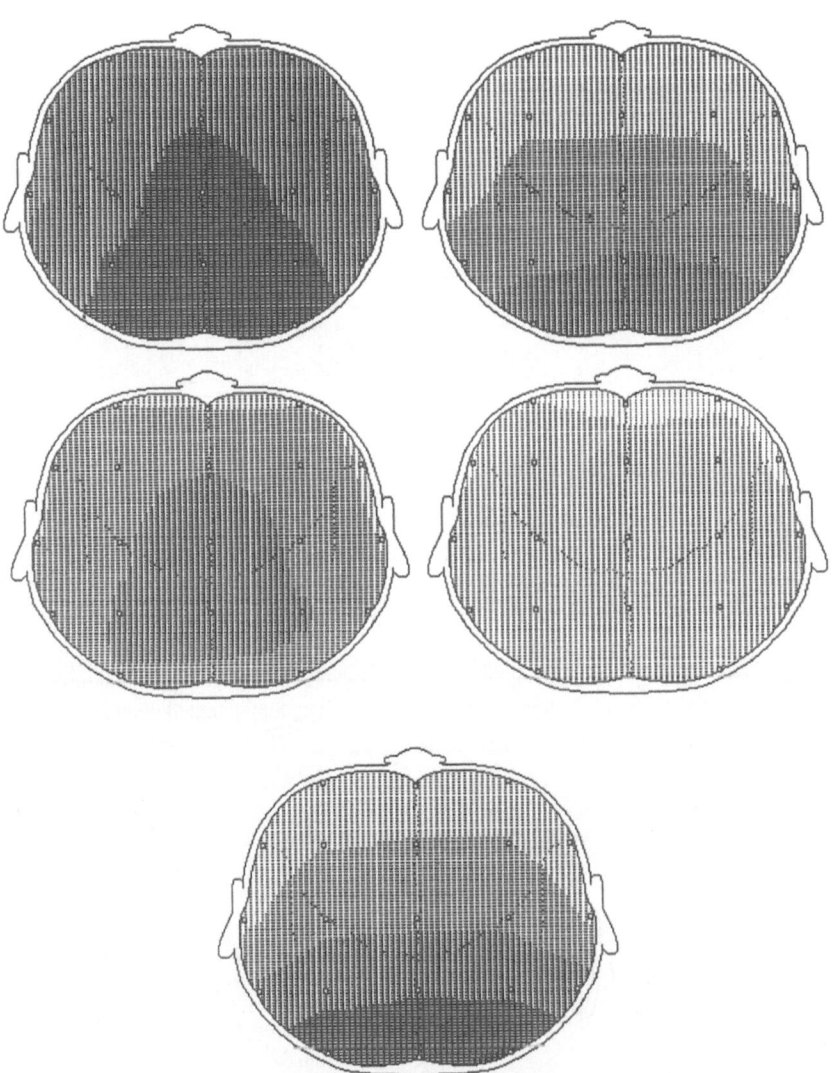

5-4e: FFT band maps (Eyes opened, 11-15 yrs).

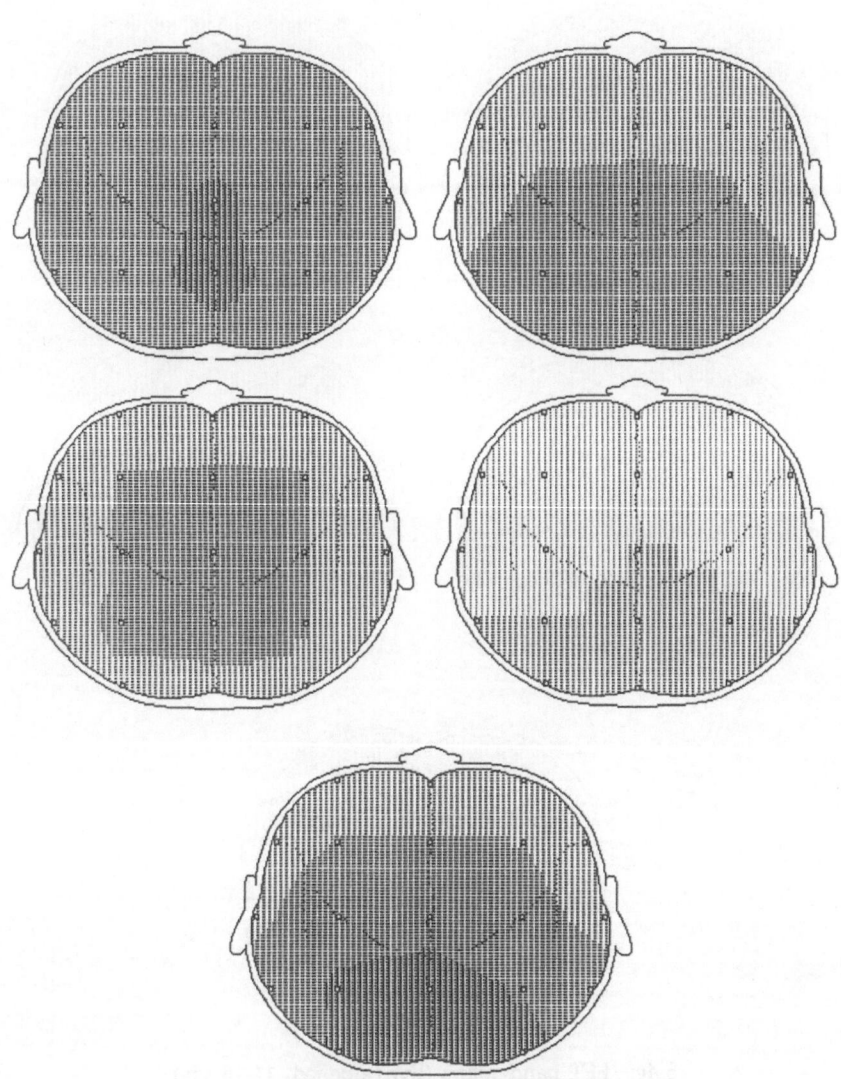

5-4f: FFT band maps (Eyes opened, 16-20 yrs).

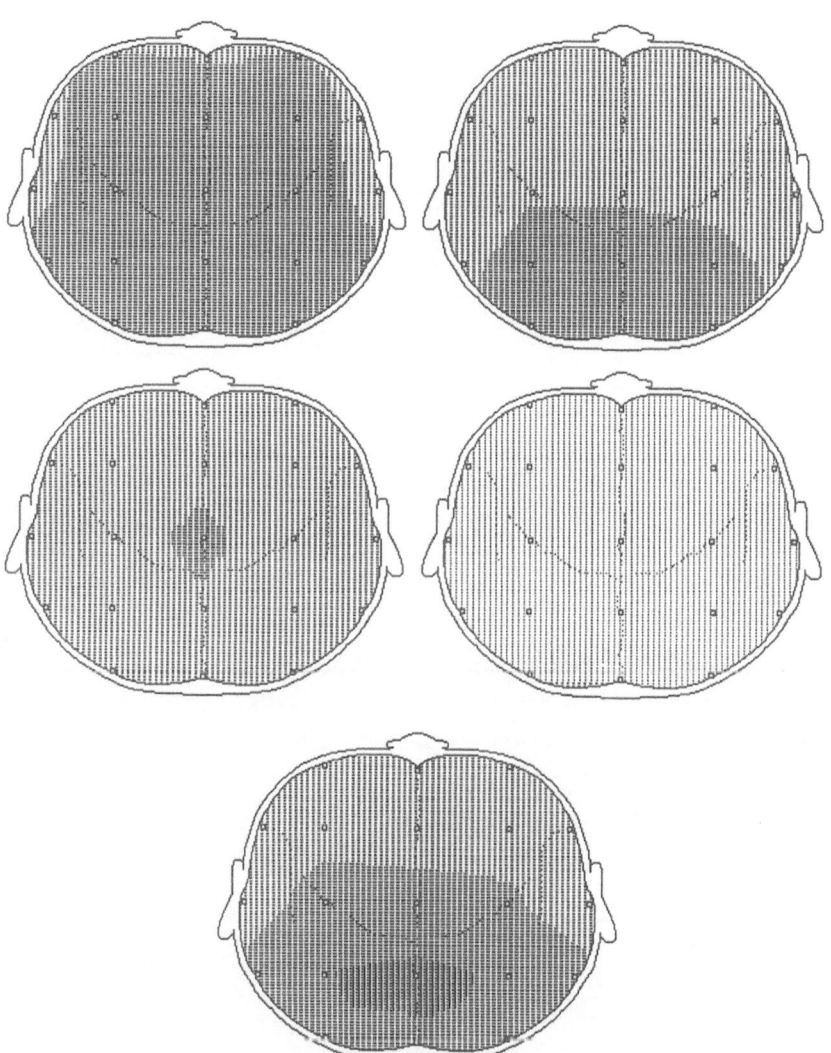

5-4g: FFT band maps (Eyes opened, 21-30 yrs).

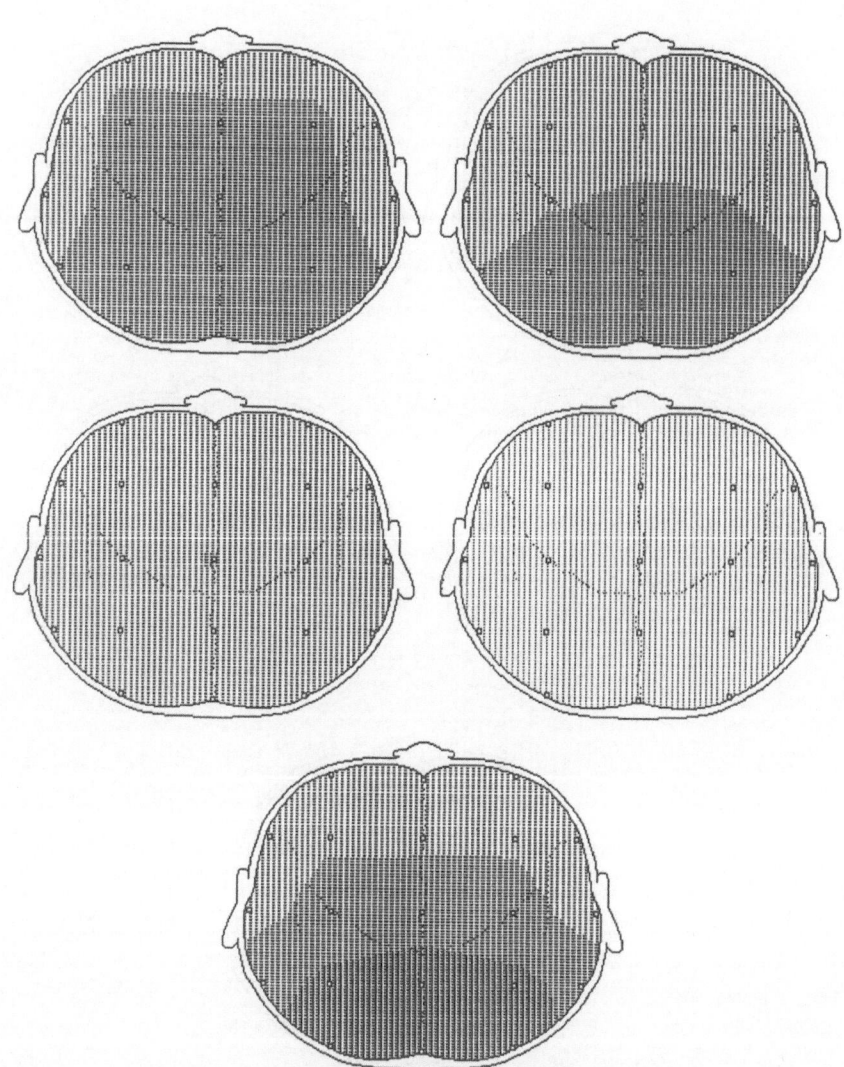

5-4h: FFT band maps (Eyes opened, 31-40 yrs).

Resting EEG

For each age group, the spectral maps displayed the total summated activity within each frequency band, and also the distribution at the peak alpha frequency. The delta and theta maps tended to be symmetrical and posterior, of medium amplitude. Alpha tended to be of highest amplitude, better in the eyes-close state as expected. Good symmetry and occipital dominance was noted. The same pattern was preserved in the eyes-open state. Beta was of low amplitude throughout, and not well-defined.

The peak alpha map was the most impressive representation, having good agreement with the alpha band maps for individual groups.

REFERENCES

Achim, A., Richer, F. and Saint-Hilaire, J.M. Methods for separating temporally overlapping sources of neuroelectric data. Brain Topography, 1988, 1: 22-28.

Adachi-Usami, E. and Lehmann, D. Monocular and binocular evoked average potential field topography: upper and lower hemiretinal stimuli. Exp. Brain Res., 1983, 50: 341-346.

Adey, W.R., Dunlop, C.W. and Hendrix, C.E. Hippocampal slow waves. AMA Arch. Neurology, 1960, 3: 74-90.

Ahn, S.S., Jordan, S.E., Nuwer, M.R., Marcus, D.R. and Moore, W.S. Computed electroencephalographic topographic brain mapping. A new and accurate monitor of cerebral circulation and function for patients having carotid endarterectomy. J. Vasc. Surg., 8: 247-254.

Alain, C., Richer, F., Achim, A. and Saint Hilaire, J. Human intracerebral potentials associated with target novel, and omitted auditory stimuli. Brain Topography, 1989, 1: 237-245.

Albert, M.L. and Helm-Estabrooks, N. Diagnosis and treatment of aphasia. Part I., JAMA, 1988, 259: 1043-1047.

Alexander, M.P. and Naeser, M.A. Cortical-subcortical differences in aphasia. Res. Publ. Assoc. Res. Nerv. Ment. Dis., 1988, 66: 215-228.

Allison, T., Matsumiya, Y., Goff, G.D. and Goff, W.R. The scalp topography of human visual evoked potentials. Electroenceph. Clin. Neurophysiol., 1977, 42: 185-197.

Amador, A.A., Sosa, P.A.V., Marqui, R.D.P., Garcia, L.G., Lirio, R.B. and Bayard, J.B. On the structure of EEG development. Electroenceph. Clin. Neurophysiol., 1989, 73: 10-19.

Anscombe F.J. "Rejection of outliers". Technometrics, 1960, 2: 123-147.

Anderer, P., Saletu, B., Kinsperger, K. and Semlitsch, H. Topographic brain mapping of EEG in neuropsychopharmacology Part I. Methodological aspects. Methods Find Exp. Clin. Pharmacol., 1987, 9: 371-84.

Anderson, R. A., Snyder, R. L. and M. M. Merzenich. The topographic organization of cortico-collicular projection from physiologically identified loci in the AI, AII, and anterior auditory cortical fields of the cat. J. Comp. Neurol., 1980, 191: 479-494.

Agianakis, G. and Anninos, P. A. Localization of epileptiform foci by means of MEG measurements. Internat. J. Neurosci., 1988, 38: 141-149.

Ashida, H., Tatsuno, J., Okamoto, J. and Maru, E. Field mapping of EEG by unbiased polynomial interpolation. Comput. Biomed. Res., 1984, 17: 267-276.

Atwick, B.A. Applied concepts in computer graphics. Prentice-Hall, New Jersey, 1984.

Bancaud, J., Talairach, J., Morel, P., Bresson, M., Bonis, A., Geier, S., Hemon, E. and Buser, P. "Generalized" epileptic seizure elicited by electrical stimulation of the frontal lobe in man. Electroenceph. Clin. Neurophysiol., 1974, 37: 275-282.

Baran, J.A., Long, R.R., Musiek, F.E. and Ommaya, A. Topographic mapping of brain electrical-activity in the assessment of central auditory nervous-system pathology. Am. J. Otology, 1988, 9: 72-76.

Barnett V. and Lewis T. "Outliers in statistical data" John Wiley and Sons, New York, N.Y., 1978.

Bartel, P.H. and Bartels, H.G. Classification strategies for topographic mapping data. In: F.H. Duffy (Eds.), Topographic Mapping of Brain Electrical Activity. Butterworth Publishers, Massachuesetts, 1986: 225-253.

Barth, D. S., Sutherling, W., Engel, J. Jr. and Beatty, J. Neuromagnetic evidence of spatially distributed generators underlying epileptiform spikes in the human brain. Science, 1984, 223: 293-296.

Barth, D. S., Sutherling, W. and Beatty, J. Fast and slow magnetic phenomena in focal epileptic seizures. Science, 1984, 226: 855-857.

Barth, D. S., Sutherling, W., Engel, J. Jr. and Beatty, J. Neuromagnetic localization of epileptiform spike activity in the human brain. Science, 1982, 218: 891-894.

Beaubernard, C., Minot, R. and Macher, J.P. Lithium clinical study by brain electrical activity mapping a case report. Pharmacopsychiatry, 1987, 20: 197-202.

Bechtereva, N.P., Vvedenskai, I.V., Dubikaitis, Y.V., Stepanova, T.S., Ovnatanov, B.S. and Usov, V.V. Localization of focal brain lesions by electroencephalography. Electroenceph. Clin. Neurophysiol., 1963, 15: 177-196.

Bencivenga, R., Wong, P.K.H., Woo, S. and Jan, J.E. Quantitative VEP analysis in children with cortical visual impairment. Brain Topography, 1989, 1: 193-198.

Berger, M.S., Kincaid, J., Ojemann, G.A. and Lettich, E. Brain mapping techniques to maximize resection, safety, and seizure control in children with brain tumors. Neurosurgery, 1989, 25: 786-792.

Bertrand, G., Olivier, A. and Thompson, C.J. Computer display of stereotaxic brain maps and probe tracts. Acta. Neurochir. (Wien), 1974, Supp. 21: 235-243.

Bertrand, O., Perrin, F. and Pernier, J. A theoretical justification of the average reference in topographic evoked potential studies. Electroenceph. Clin. Neurophysiol., 1985, 62: 462-464.

Bick, C.H. An EEG-mapping study of laughing - coherence and brain dominances. International J. Neurosc., 1989, 47: 31-40.

Bickel P.J. and Doksum K.A. "Mathematical statistics" Holden-Day, Oakland, CA., 1977.

Bickford, R.G. EEG color mapping (Toposcopy) - advantages and pitfalls, clinical and research perspectives. Amer. J. EEG Tech., 1989, 29: 19-28.

Bickford, R.G. A combined EEG and evoked potential procedure in clinical EEG (automated cerebral electrogram - ace test). In: N. Yamaguchi and K. Fujisawa (Eds.), Recent Advances in EEG and EMG Data Processing. Elsevier, 1981, 217-260.

Bickford, G.E. Computer analysis of Background Activity. In: A. Remond (Ed.), EEG Informatics. A Didactic Review of Methods and Applications of EEG Data Processing. Elsevier, New York, 1977, 215-232.

Borda, R. P. The 40/sec middle latency response in Alzheimer's disease, Parkinson's disease, and age-matched controls. Unpublished doctoral dissertation, University of Texas, 1983.

Borg, E., Spens, K.E. and Tonnquist, I. Auditory brain map, effects of age. Scand. Audiol., 1988, Supp. 30: 161- 164.

Borg, E., Spens, K.E., Tonnquist, I. and Rosen, S. Brain map. New possibilities in diagnosis of central auditory disorders?   Acta. Otolaryngol., (Stockh), May-June 1, 103: 612.

Bostem, F. and Degossely, M. Spectral analysis of alpha rhythm during schultz's autogenic training. Electroenceph. Clin. Neurophysiol., 1978, Supp. 34: 181-190.

Bourne, J.R. Childers, D.G. and Perry, N.W., Jr. Topological characteristics of the visual evoked response in man. Electroenceph. Clin. Neurophysiol., 1971, 30: 423-436.

Bradshaw, J.L., Burden, V. and Nettleton, N.C. Dichotic and dichhaptic techniques. Neuropsychologia., 1986, 24: 79-90.

Brandies, D. and Lehmann, D. Event-related potentials of the brain and cognitive processing: approaches and applications. Neuropsychologica., 1986, 24: 151-168.

Brazier, M.A. Cerebral localization. The search for functional representation in the cortex. Acta. Neurobiol. Exp. (Warsz), 1975, 35: 529-35.

Brazier, M.A.B. Computer techniques in EEG analysis. Electroenceph. Clin. Neurophysiol., 1961, Supp. 20: 1-98.

Breiman L., Friedman J.H., Olshen R.A. and Stone C.J. "Classification and regression trees" Wadsworth, Belmont, CA., 1984.

Breitling, D., Guenther, W. and Rondot, P. Auditory - perception of music measured by brain electrical-activity mapping. Neuropsychologia., 1987, 25: 765-774.

Brenner, D., Kaufman, L. and Williamson, S. J. Somatically evoked fields of the human brain. Science, 1978, 199: 81-83.

Brenner, D., Okada, Y., Maclin, E., Williamson, S. J. and Kaufman, L. Evoked magnetic fields reveal different visual areas in the human cortex. In: S.N. Erne', H.D. Hahlbohm and H. Lubbig (Eds.), Biomagnetism, Walter de Gruyter, Berlin, 1981: 431-444.

Breslau, J., Starr, A., Sicotte, N., Higa, J. and Buchsbaum, M.S. Topographic EEG changes with normal aging and SDAT. Electroenceph. Clin. Neurophysiol., 1989, 72: 281-289.

Brickett, P., Robertson, A., Crisp, D. and Weinberg, H. Comparison of the magnetic fields related to alpha activity and visual evoked responses. EPIC VIII, Stanford California, 1986.

Brigham, E.O. The fast fourier transform. Prentice-Hall, New Jersey, 1974.

Brown, W.S. and Lehmann, D. Verb and noun meaning of homophone words activate different cortical generators: a topographical study of evoked potential fields. Exp. Brain Res., 1979, Supp. 2: 159-168.

Buchsbaum, M.S., Awsare, S.V., Holcomp, H.H., Delisi, L.E., Hazlett, E., Carpenter, W.T., Pickar, D. and Morihisa, J.M. Topographic differences between normals and schizophrenics: the N120 evoked potential component., Neuropsychobiol., 1986, 15: 1-6.

Buchsbaum, M.S., Cappelletti, Coppola, R., Regal, F., King, A.C. and Van Kammen, D.P. New Methods to determine the CNS effects of antigeriatric compounds: EEG topography and glucose use. Drug Develop. Rev., 1982, 2: 489-496.

Buchsbaum, M.S., Coppola, R. and Cappelletti, J. Positron emission tomography, EEG and evoked potential topography: new approaches to local function in pharmaco - encephalography. Hermann, 1982.

Buchsbaum, M.S., Coppola, R., Gershon, E.S., vanKammen, D. P. and Nurnberger, J.I. Evoked potential measures of attention and psychopathology. Adv. Biol. Psychiat., 1981, 6: 186-194.

Buchsbaum, M.S., King, A.C., Cappelletti, J., Coppolla, R. and vanKammen, D.P. Visual evoked potential topography in patients with schizophrenia and normal controls. Adv. Biol. Psychiat., 1982, 9: 50-56.

Buchsbaum, M.S., Lee, S.H., Haier, R., Wu, J.C., Green, M. and Tang, S.W. Effects of amoxapine and imipramine on evoked-potentials in the continuous performance - test in patients with affective-disorder. Neuropsychobiol., 1988, 20: 15-22.

Buchsbaum, M.S., Mendelson, W.B., Duncan, W.C., Coppola, R., Kelsoe, J. and Gillin, J.C. Topographic cortical mapping of EEG sleep stages during daytime naps in normal subjects. SLP, 1982, 5: 248-255.

Buchsbaum, M.S., Rigal, F., Coppola, R., Cappelletti, J., King, C. and Johnson, J. A new system for gray-level surface distribution maps of electrical activity. Electroenceph. Clin. Neurophysiol., 1982, 53: 237-242.

Carelli, P., Modena, I., Ricci, G. B. and Romani, G.L. Magneto-encephalography. In: S.J. Williamson, G.L. Romani, L. Kaufman and I. Modena (Eds.), Biomagnetism: An Interdisciplinary Approach. NATO ASI Series A. Plenum Press, New York, 1983: 469-482.

Celesia, G.G. and Puletti, F. Auditory cortical areas of man. Neurol., 1969, 9: 211-20.

Chapman, R. M., Ilmoniemi, R. J., Barbanera, S. and Romani, G. L. Selective localization of alpha brain activity with neuromagnetic measurements. Electroenceph. Clin. Neurophysiol., 1984, 58: 569-572.

Chauvel, P., Buser, P., Badier, J.M., Liegeois-Chauvel, C., Marquis, P. and Bancaud, J. The "epileptogenic zone" in humans: representation of intercritical events by spatio - temporal maps (French). Rev. Neurol., (Paris), 1987, 143: 443-450.

Chen, A.C., Dworkin, S.F., Haug, J. and Gehrig, J. Topographic brain measures of human pain and pain responsivity. Pain, 1989, 37: 129-141.

Cheyne, D. O. Magnetic and electric field measurements of brain activity preceding voluntary movements: Implications for supplementary motor area function. Ph.D. Thesis, Simon Fraser University, 1987.

Cheyne, D. and Weinberg, H. Neuromagnetic fields accompanying unilateral finger movements: pre-movement and movement-evoked fields. Experimental Brain Research, 1989, 78: 604-612.

Childers, D.G. Automated Visual evoked-response system. Med. and Biol. Eng. and Comput., 1977, 15: 374-380.

Clark, S.A., Allard, T., Jenkins, W.M. and Merzenich, M.M. Receptive fields in the body-surface map in adult cortex defined by temporally correlated inputs. Nature, 1988, 332: 444-445.

Cleveland, W.S. "The elements of graphing data" Wadsworth, Monterey, CA., 1985.

Coburn, K.L. and Moreno, M.A. Facts and artifacts in brain electrical activity mapping. Brain Topography, 1988, 1: 37-45.

Coburn, K.L., Sullivan, C.H. and Hundley, J. High-tech maps of the brain. Am. J. Nurs., 1988, 88: 1500-1501.

Cochran W.G. "Sampling techniques" 3rd edition, John Wiley and Sons, New York, N,Y, 1977.

Cohen, L.G. and Hallett, M. Methodology for non-invasive mapping of human motor cortex with electrical stimulation. Electroenceph. Clin. Neurophysiol., 1988a, 69: 403-411.

Cohen, L.G. and Hallett, M. Noninvasive mapping of human motor cortex. Neurol., 1988b, 38: 904-909.

Cohen, D. Evidence of magnetic fields produced by alpha-rhythm currents. Science, 1968, 161: 784.

Cohen, D. Magnetoencephalography: detection of the brain's electrical activity with a superconducting magnetometer. Science, 1972, 175: 664-666.

Cohen, D. and Cuffin, B. N. Search for MEG signals due to auditory brain-stem stimulation. In: H. Weinberg, G. Stroink and T. Katila (Eds.), Biomagnetism: Applications and Theory, Pergamon Press, New York, 1985, 316-320.

Cohen, D., Cuffin, B. N., Kennedy, J. G., Lombroso, C. T., Gumnit, R. J. and Schomer, D. L. Comparison of MEG versus EEG spike localization: Some results in a patient group with focal seizures. Poster presented at the Annual meeting of the American Epilepsy Society, San Francisco, California, October, 1988.

Cohen, D., Edelsack, E. A. and Zimmerman, J. E. Magnetocardiograms taken inside a shielded room with a superconducting point-contact magnetometer. Appl. Phys. Letters, 1970, 16: 278-280.

Comacchio, F., Grandori, F., Magnavita, V. and Martini, A. Topographic brain mapping of middle latency auditory evoked potentials in normal subjects, Scand. Audiol., 1988, Supp. 30: 165-172.

Coppola, R., Topographic methods of functional cerebral analysis, FEHC, 1982, 71-78.

Coppola, R., Buchsbaum, M.S. and Rigal, F. Computer generation of surface distribution maps of measures of brain activity. Comput. Biol. Med., 12: 1982.

Coppola, R. and Morgan, N.T. Multi-channel amplifier system for computerized EEG analysis. Electroenceph. Clin. Neurophysiol., 1987, 87: 8-10.

Cowey, A. Cortical maps and visual perception: the Grindley memorial lecture. Q. J. Exp. Psychol., 1979, 31: 1-17.

Crisp, D. Neuromagnetic localization of current dipole generators in complex partial epilepsy. M.A. thesis, Simon Fraser University, 1986.

Cuffin, N.B. Effects of inhomogeneous regions on electric potentials and magnetic fields: two special cases. J. Appl. Phys., 1982, 53: 9192-9197.

Damasio, A.R. Neuroscience and cognitive science in the study of language and the brain. Res. Publ. Assoc. Res. Nerv. Ment. Dis., 1988, 66: 275-282.

Darcey, T.M., Ary, J.P. and Fender, D.H. Spatio-temporal visually evoked scalp potentials in response to partial- field patterned stimulation. Electroenceph. Clin. Neurophysiol., 1980, 50: 348-355.

DeArmond, S., Fusco, M. and Dewey, M. Structure of the human brain. Second Edition. Oxford Press, New York, 1976.

Deecke, L., Boschert, J., Brickett, P. and Weinberg, H. Magnetoencephalographic evidence for possible supplementary motor area participation in human voluntary movement. In: H. Weinberg, G. Stroink and T. Katila (Eds.), Biomagnetism: Applications and Theory, Pergamon Press, New York, 1985: 369-372.

Deecke, L., Boschert, J., Weinberg, H. and Brickett, P. Magnetic fields of the human brain (Bereitschaftsmagnetfeld) preceding voluntary foot and toe movements. Exp. Brain Res., 1983, 52: 81-86.

Deecke, L., Weinberg, H. and Brickett, P. Magnetic fields of the human brain accompanying voluntary movement: Bereitschaftsmagnetfeld. Exp. Brain Res., 1982, 48: 144-148.

Deiber, M.P., Giard, M.H. and Mauguiere, F. Separate generators with distinct orientations for N20 and P22 somatosensory evoked potentials to finger stimulation?, Electroenceph. Clin. Neurophysiol., 1986, 65: 321-334.

Deiber, M.P., Ibanez, V., Fischer, C., Perrin, F. and Mauguiere, F. Sequential mapping favours the hypothesis of distinct generators for Na and Pa middle latency auditory evoked potentials. Electroenceph. Clin. Neurophysiol., 1988, 71: 187-197.

Della Sala, S. and Spinnler, H., Anton's (Anton-Babinski's) syndrome associated with Dide-Botcazo's syndrome: a case report of denial of cortical blindness and amnesia. Schweiz. Arch. Neurol. Psychiatr., 1988, 139: 5-15.

Desmedt, J.E. and Bourguet, M. Color imaging of parietal and frontal somatosensory potential fields evoked by stimulation of median or posterior tribal nerve in man. Electroenceph. Clin. Neurophysiol., 1985, 62: 1-17.

Desmedt, J.E. and Chalklin, V. New method for titrating differences in scalp topographic patterns in brain evoked potential mapping. Electroenceph. Clin. Neurophysiol., 1989, 74: 359-366.

Desmedt, J.E. and Cheron, G. Central somatosensory conduction in man: neural generators and interpeak latencies of the far-field components recorded from neck and right or left scalp and earlobes. Electroenceph. Clin. Neurophysiol., 1980a, 50: 382-403.

Desmedt, J.E. and Cheron, G. Somatosensory evoked potentials to finger stimulation in healthy octogenarians and in young adults: wave forms, scalp topography and transit time of partial and frontal components, Electroenceph. Clin. Neurophysiol., 1980b, 50: 404-425.

Desmedt, J.E. and Nguyen, T.H. Bit-mapped colour imaging of the potential fields of propagated and segmental subcortical components of somatosensory evoked potentials in man. Electroenceph. Clin. Neurophysiol., 1984, 58: 481-497.

Desmedt, J.E., Nguyen, T.H. and Bourguet, M. Bit-mapped color imaging of human evoked potentials with reference to the N20, P22, P27 and N30 somatosensory responses. Electroenceph. Clin. Neurophysiol., 1987, 68: 1-19.

Desmedt, J.E. and Tomberg, C. Mapping early somatosensory evoked potentials in selective attention: critical evaluation of control conditions used for titrating by difference the cognitive P30, P40, P100 and N140., Electroenceph. Clin. Neurophysiol., 1989, 74: 321-346.

Dillon W.R. and Goldstein M. "Multivariate analysis, methods and applications" John Wiley and Sons, New York, N.Y., 1984.

Dierks, T., Maurer, K. and Zacher, A. Brain mapping of EEG in autogenic training (AT). Psychiatry Research, 1989, 29: 433-434.

Dimond, S.J. Brain circuits for consciousness. Brain Behav. Evol., 1976: 376-395.

Dolce, G. and Waldeier, H. Spectral and multivariate analysis of EEG changes during mental activity in man. Electroenceph. Clin. Neurophysiol., 1974, 36: 577-584.

Dondey, M. Transverse topographical analysis of petit mal discharges: diagnostical and pathogenic implications. Electroenceph. Clin. Neurophysiol., 1984, 58: 481-497.

Drake, M.E. and Shy, K.E., Predictive value of electroencephalography for electroconvulsive-therapy. Clin. Electroenceph. 1989, 20: 55-57.

Draper N.R. and Smith H. "Applied regression analysis" 2nd Edition, John Wiley and Sons, New York, N.Y., 1981.

Drasdo, N. and Furlong, P. Coordinate systems for evoked potential topography. Electroenceph. Clin. Neurophysiol., 1988, 71: 469-473.

Dubinsky, J. and Barlow, J.S. A simple dot-density topogram for EEG. Electroenceph. Clin. Neurophysiol., 1980, 48: 473-477.

Ducati, A., Fava, E. and Motti, E.D. Neuronal generators of the visual evoked potentials: intracerebral recording in awake humans. Electroenceph. Clin. Neurophysiol., 1988, 71: 89-99.

Duff, T.A. Topography of scalp recorded potentials evoked by stimulation of the Digits. Electroenceph. Clin. Neurophysiol., 1980a, 49: 452-460.

Duff, T.A. Multichannel topographic analysis of human somatosensory evoked potentials. PCN, 1980b, 7: 69-86.

Duffy, F.H. Brain electrical activity mapping (BEAM): computerized access to complex brain function. Int. J. Neurosci., 1981, 13: 55-65.

Duffy, F.H. Topographic display of evoked potentials: clinical applications of brain electrical activity mapping (BEAM). Ann. N.Y. Acad. Sci., 1982, 388: 183-196.

Duffy, F.H. Brain electrical activity mapping - clinical applications, Psychiatry Research, 1989, 29: 379-384.

Duffy, F.H., Albert, M.S. and McAnulty, G. Brain electrical activity in patients with presenile and senile dementia of the Alzheimer type. Annals of Neurol. 1984, 16: 439-448.

Duffy, F.H., Albert, M.S., McAnulty, G. and Garvey, A.J. Age-related differences in brain electrical activity of healthy subjects. Annals of Neurol., 1984, 16: 430-438.

Duffy, F.H., Bartels, P.H. and Burchfiel, J.L. Significance probability mapping: an aid in the topographic analysis of brain electrical activity. Electroenceph. Clin. Neurophysiol., 1981, 51: 455-462.

Duffy, F.H., Burchfiel, J.L., and Lombroso, C.T.D. Brain electrical activity mapping (BEAM): a method for extending the clinical utility of EEG and evoked potential data. Annals of Neurol., 1979, 5: 309-321.

Duffy, F.H., Denckla, M.B., Bartels, P.H. and Sandini, G. Dyslexia: regional differences in brain electrical activity by topographic mapping. Annals of Neurol., 1980, 7: 412-420.

Duffy, F.H., Denckla, M.B., Bartels, P.H., Sandini, G. and Kiessling, L.S. Dyslexia: automated diagnosis by computerized classification of brain electrical activity. Annals of Neurol., 1980, 7: 421-428.

Duffy, F.H., Denckla, M.B., McAnulty, G.B. and Holmes, J. A. Neurophysiological studies in dyslexia. Res. Publ. Assoc. Res. Nerv. Ment. Dis. 1988, 66: 149-170.

Duffy, F.H., Jensen, F., Erba, G., Burchfiel, J.L. and Lombroso, C.T. Extration of clinical information from electroencephalographic background activity: the combined use of brain electrical activity mapping and intravenous sodium thiopental. Annals of Neurol., 1984, 15: 22-30.

Dumermuth, G. Fundamentals of spectral analysis in electroencephalography. In: A. Remond (Eds.), EEG informatics: A Didactic Review of Methods and Applications of EEG Data Processing. Elsevier/North-Holland Biomedical Press, Amsterdam, 1977, 83-106.

Edwards, L. and Drasdo, N. Scalp distribution of visual evoked potentials to foveal pattern and luminance stimuli. Documenta Ophthalmologica, 1987, 66: 301-311.

Efron, B. The jackknife, the bootstrap and other resampling plans. Society for Industrial and Applied Mathematics, Philadelphia, 1982.

Epstein, C.M. and Brickley, G.P. Interelectrode distance and amplitude of the scalp EEG. Electroenceph. Clin. Neurophysiol., 1985, 60: 287-292.

Estrin, T. and Uzgalis, R. Computerized display of spatio-temporal EEG patterns, IEEET, 1969, 16(3).

Etevenon, P., Bertaut, A., Mitermite, F., Eustache, F., Lespaisant, J., Lechevalier, B. and Zarifian, E. Inter- and intra-individual probability maps in EEG cartography by use of nonparametric fisher tests. Brain Topography, 1989, 2: 81-89.

Etevenon, P. and Gaches, J. Quantitative EEG maps in neuro-psychiatry: problems and perspectives. Clin. Neurpharmacol., 1984, 1: 122-123.

Etevenon, P., Peron-Magnan, P., Gueguen, B., Ghanem, M., Gaches, J. and Deniker, P. Value of quantitative EEG and EEG mapping in medicine (French). Ann. Med. Interne (Paris), 1987, 138: 13.

Etevenon, P., Tortrat, D. and Benkt, C. EEG cartography by means of statistical group studies - activation by visual attention. Neuropsychobiol., 1985, 13: 141-146.

Faienza, C., Capone, C., Sani, E., Villani, D. and Prati, G. EEG mapping in a child with RETT syndrome. Psychiatry Research, 1989, 29: 425-426.

Fallside, F., Woods, W.A. (Eds.), Computer speech processing. Prentice-Hall, New Jersey, 1985.

Faux, S.F., Shenton, M.E., McCarley, R.W., Torello, M.W. and Duffy, F.H. Differentiation of schizophrenics and normal controls is enhanced by the Goodin subtraction procedure. International J. Neurosc. 1988, 39: 117.

Faux, S.F., Torello, M.W., McCarley, R.W., Shenton, M.E. and Duffy, F.H. P300 in schizophrenia: confirmation and statistical validation of temporal region deficit in P300 topography. Biol. Psychiatry, 1988, 23: 776-790.

Fender, D. Source localization of brain electrical activity. In: A.S. Gevins and A. Remond (Eds.), Methods of analysis of brain electrical and magnetic signals. Handbook of EEG and Clinical Neurophysiology Elsevier Science Publishers, Amsterdam, 1987: (Revised series) 1: 355-403.

Fender, D.H. Source localization of brain electrical activity. In: A.S. Gevins and A. Remond (Eds.), Methods of Analysis of Brain Electrical and Magnetic Signals. Handbook of Electroencephalography and Clinical Neurophysiology. Elsevier, Revised Series, 1987, 1: 355-403.

Findji, F., Catani, P. and Liard, C. Topographical distribution of Delta rhythms during sleep: evolution with age. Electroenceph. Clin. Neurophysiol., 1981, 51: 659-665.

Fisch, B.J. and Pedley, T.A. The role of quantitative topographic mapping or neurometrics in the diagnosis of psychiatric and neurological disorders - the cons. Electroenceph. Clin. Neurophysiol., 1989, 73: 5-9.

Fisch, B.J., Pedley, T.A. and Keller, D.L. A topographic background symmetry display for comparison with routine EEG. Electroenceph. Clin. Neurophysiol., 1988, 69: 491-494.

Fiumara, R., Campitelli, F., Romani, G. L., Leoni, R., Caporali, M., Zanasi, M., Cappiello, A., Fioriti, G. and Modena, I. Neuromagnetic study of endogenous fields related to the contingent negative variation. In: H. Weinberg, G. Stroink, T. Katila (Eds.), Biomagnetism: Applications and Theory, Pergamon Press, New York, 1985, 336-342.

Flynn, J. and Boder, E. Clinical and electrophysiological correlates of dysphonetic and dyseidetic dyslexia. In: J.F. Stein (Ed.), Vision and dyslexia, U. of Oxford Press, Oxford, in press.

Flynn, J. and Deering, W. Subtypes of dyslexia: investigation of Boder's system using quantitative neurophysiology. Developmental Medicine and Child Neurol., 1989, 31: 215-223.

Flynn, J.M. and Deering, W.M. Topographic brain mapping and evaluation of dyslexic-children, Psychiatry Research, 1989, 29: 407-408.

Freeman, W.J. and Maurer, K. Images and imaginings from computerized brains. Psychiatry Research, 1989, 29: 239-245.

Friedman, L.R. Unwrapping the riddle of the brain-injured patient by utilizing the BEAM EEG. Am. J. Forensic Psych., 1982, 3: 467-451.

Fuenfgeld, E.W. Brain electrical-activiey mapping in different stages of SDAT, Psychiatry Research, 1989, 29: 411-412.

Galaburda, A.M. The pathogenesis of childhood dyslexia. Res. Publ. Assoc. Res. Nerv. Ment. Dis., 1988, 66: 127-137.

Galambos, R., Makeig, S. and Talmachoff, P. J. A 40-Hz auditory potential recorded from the human scalp. Proc. Nat. Acad. Sci., 1981, 78: 2643-2647.

Garber, H.J., Weilburg, J.B., Duffy, F.H. and Manschreck, T.C., Clinical use of topographic brain electrical activity mapping in Psychiatry, J. Clin. Psychiatry, 1989, 50: 205-211.

Gasser, T., Jennen-Steinmetz, C., Sroka, L., Verleger, R. and Mocks, J. Development of the EEG of school-age children and adolescents. II. Topography, Electroenceph. Clin. Neurophysiol., 1988, 69: 100-109.

Geselowitz, D.B. On bioelectric potentials in an inhomogeneous volume conductor. J. Biophys., 1967, 7: 1-11.

Geschwind, N., Galaburda, A. and LeMay, M. Morphological and physiological substrates of language and cognitive development. Res. Publ. Assoc. Res. Nerv. Ment. Dis., 1979, 57: 31-41.

Gevins, A.S. Advances in measuring higher brain functions, Parts 1 and 2. J. Clin. Neurophysiol., 1988, 5: 325.

Gevins, A.S. Overview of computer analysis. In: A.S. Gevins and A. Remond (Eds.), Methods of Analysis of Brain Electrical and Magnetic Signals. Handbook of Electroencephalography and Clinical Neurophysiology. Revised Series, 1987, 1: 31-83.

Gevins, A.S. and Bressler, S.L. Functional topography of the human brain. In: G. Pfurtscheller and F. Lopes da Silva (Eds.), Functional Brain Imaging. Hans Huber Publishers, Toronto, 1988: 99-116.

Gevins, A., Morgan, N.H., Bressler, S.L., Cutillo, B.A., White, R.M., Illes, J., Greer, D.S. and Doyle, J.C. Human Neuroelectric patterns predict performance accuracy. Science, 1987, 235: 580-585.

Gevins, A. Handbook of EEG and clinical neurophysiology revised series, 1987, Chp. 3: 31-83.

Gevins, A. Dynamic functional topography of cognitive tasks. Brain Topography. 1989, 2: 37-56.

Ghez, C. and Vicario, D. Discharge of red nucleus neurons during voluntary muscle contraction: activity patterns and correlations with isometric force. J. Physiol. (Paris), 1974, 74: 283-285.

Ghez, C. Voluntary movment. In: E.R. Kandel and J.H. Schwartz (Eds.), Principles of Neural Sciences, Elsiever, New York, 1985, 487-501.

Giard, M.H., Perrin, F., Pernier, J. and Peronnet, F. Several attention-related wave forms in auditory areas: a topographic study. Electroenceph. Clin. Neurophysiol., 1988, 69: 371-384.

Giorgi, C., Cerchiari, U., Broggi, G., Birk, P. and Struppeler, A. Digital image processing to handle neuroanatomical information and neurophysiological data. Appl. Neurophysiol., 1985, 48: 30-33.

Gloor, P. Neuronal generators and the problem of localization in electroencephalography: application of volume conductor theory to electroencephalography. J. Clin. Neurophysiol., 1985, 2: 327-354.

Gnanadesikan R. "Methods for statistical data analysis of multivariate observations". John Wiley and Sons, New York, N.Y., 1977.

Goff, G.D. Matsumiya, Y., Allison, T. and Goff, W.R. The scalp topography of human somatosensory and auditory evoked potentials. Electroenceph. Clin. Neurophysiol., 1977, 42: 57-76.

Goff, W., Allison, T., Williamson, P. and Van Gilder, J. Scalp topography in the localization of intracranial evoked potential sources. In: (Ed.), Proceeding of the 4th International Congress on Event-Related Slow Potentials of the Brain, US Government Printing Office, Washington, DC, 1978, 526-632.

Goff, W.R., Rosner, B.S. and Allison, T. Distribution of cerebral somatosensory evoked responses in normal man. Electroenceph. Clin. Neurophysiol., 1962, 14: 697-713.

Golden, G.S. Neurobiological correlates of learning disabilities. Ann. Neurol., 1982, 12(5).

Goldman, S., Vivian, W.E., Chien, C.K. and Bowes, H.N. Electronic mapping of the activity of the heart and the brain. Science, 1948, 108: 720-723.

Goldring, S. and Ratcheson, R. Human motor cortex: sensory input from single neuron recordings. Science, 1972, 175: 1493-1495.

Gotman, J. and Gloor, P. Automatic recognition and quantification of interictal epileptic activity in the human scalp EEG. Electroenceph. Clin. Neurophysiol., 1976, 41: 513-529.

Gotman, J., Ives, J.R. and Gloor, P. Frequency content of EEG and EMG at seizure onset: possibility of removal of EMG artefact by digital filtering. Electroenceph. Clin. Neurophysiol., 1981, 52: 626-639.

Gotman, J., Skuce, D.R., Thompson, C.J., Gloor, P., Ives, J.R. and Ray, W.F. Clinical applications of spectral analysis and extraction of features from electroencephalograms with slow waves in adult patients. Electroenceph. Clin. Neurophysiol., 1973, 35: 225-235.

Grandori, F. Field analysis of auditory evoked brainstem potentials. Hear Res, 1986, 21: 51-58.

Grass, A.M. and Gibbs, F.A. A Fourier Transform of the Electroenceph. J. Neurophysiol., 1938, 1: 521-526.

Gregory, D.L. and Wong, P.K.H. Topographical analysis of the centro-temporal discharges in benign rolandic epilepsy of childhood. Epilepsia, 1984, 25: 705-711.

Gregory, K.L., Wong, P.K.H., Farrell, K. and Ramsay, R.E. Spike topography in benign rolandic epilepsy. Epilepsia, 1984, 25: 688.

Grillon, C. and Buchsbaum, M.S. Computed EEG topography of response to visual and auditory stimuli. Electroenceph. Clin. Neurophysiol., 1986, 63: 42-53.

Grillon, C. and Buchsbaum, M.S. EEG topography of response to visual stimuli in generalized anxiety disorder. Electroenceph. Clin. Neurophysiol., 1987, 66: 337-348.

Grobstein, P. Between the retinotectal projection and directed movement: topography of a sensorimotor interface. Brain Behav. Evol., 1988, 31: 34-48.

Grunewald-Zuberbier, E. and Grunewald, G. Goal directed movement potentials of human cerebral cortex. Exp. Brain Res., 1978, 33: 135-138.

Guedes De Oliveira, P.H.H. and Lopes Da Silva, F.H. A topographical display of epileptiform transients based on a statistical approach. Electroenceph. Clin. Neurophysiol., 1980, 48: 710-714.

Guenther, W. and Breitling, D. Predominant sensorimotor area left hemisphere dysfunction in schizophrenia measured by brain electrical activity mapping. Biol. Psychiat., 1985, 20: 515-532.

Guenther, W., Breitling, D., Banquet, J.P., Marcie, P. and Rondot, P. EEG mapping of left hemisphere dysfunction during motor performance in schizophrenia. Biol. Psychiat., 1986, 21: 249-262.

Hachinski, V. Brain mapping. Archives of Neurology, 1989, 46: 1136.

Halliday, A.M., Barrett, G., Halliday, E. and Michael, W. F. The topography of the pattern-evoked potential. In: J.E. Desmedt (Ed.), Visual evoked potentials in man: New Developments, Clarendon Press, Oxford, 1977, 121-133.

Hamming, R.W. Digital filters. 2nd ed., Prentice-Hall, New Jersey, 1983.

Hari, R. Somatically evoked magnetic fields. Med. and Biol. Engineering and Computing, 1985, 22 (Supp. 1): 29-31.

Hari, R., Aittoniemi, K., Jarvinen, M. L., Katila, T. and Varpula, T. Auditory evoked transient and sustained magnetic fields of the human brain. Localization of neural generators. Exp. Brain Res., 1980, 40: 237-240.

Hari, R., Antervo, A., Katila, T., Poutanen, T., Seppanen, M., Tuomisto, T. and Varpula, T. Cerebral magnetic fields associated with voluntary limb movements in man. Il Nuovo Cimento 2D, 1983, 1: 484-495.

Hari, R., Hamalainen, M., Ilmonicmi, R., Kaukoranta, E., Reinikalnen, K., Salminen, J., Alho, K., Naatanen, R. and Sams, M. Responses of the primary auditory cortex to pitch changes in a sequence of tone pips: Neuromagnetic recordings in man. Neuroscience Letters, 50: 127-132.

Hari, R., Reinkainen, K., Kaukoranta, E., Hamalainen, M., Ilmoniemi, R., Penttinen, A., Salminen, J. and Teszner, D. Somatosensory evoked cerebral magnetic fields from SI and SII in man. Electroenceph. Clin. Neurophysiol., 1984, 57: 254-263.

Harner, R.N. and Riggio, S. Application of singular value decomposition to topographic analysis of flash-evoked potentials. Brain Topography, 1989, 2: 91-98.

Harner, R.N. Topographic analysis of multichannel EEG data. In: D. Samson-Dollfus, J.D. Guieu, J. Gotman, and P. Etevenon (Eds.), Statistics and topography in quantitative EEG. Elsevier, Amsterdam, 1988, 49-61.

Harner, R.N., Brain mapping or spatial analysis? Brain Topography, 1989, 1: 73-76.

Harner, R.N., Jackel, R.A., Mawhinney-Hee, M.R. and Sussman, N.M. Computed EEG Topography in Epilepsy, Rev. Neurol. (Paris), 1987, 143: 457-461.

Harner, R.N. and Ostergren, A.K. Computed EEG topography, Electroencephal. Clin. Neurophysiol., 1978, Supp. 34: 151-161.

Harner, R.N. and Ostergren, K.A. Computed EEG topography: A new method for the study of neurological disorders, Trans. Am. Neurol. Assn., 1978,103: 127-129.

Harris, J.A. and Bickford, R.G. Spatial display and parameter computation of the human epileptic spike focus by computer. Electroenceph. Clin. Neurophysiol., 1968, 24: 281-282.

Harris, J.A., Melby, G.M. and Bickford, R.G. Computer-controlled multidimensional display device for investigation and modeling of physiologic systems. Comput. Biom., 1969, 2: 519-536.

Harrop, R., Weinberg, H., Brickett, P., Dykstra, C., Robertson, A., Cheyne, D., Baff, M. and Crisp, D. An inverse solution method for the simultaneous localisation of two dipoles. Paper presented at the meeting of the Institute of Physics: Magnetism Subcommittee, Milton Keynes, England, 1986.

Harrop, R., Weinberg, H., Brickett, P., Dykstra, C., Robertson, A., Cheyne, D. O., Baff, M., and Crisp, D. The biomagnetic inverse problem: some theoretical and practical considerations. Physics in Medicine and Biology, 1987, 32(12): 1545-1557.

Hendee, W.R. Dyslexia, JAMA, 1989, 261: 2236-2239.

Hjorth, B. Eigenvectors and eigenfunctions in spatiotemporal EEG analysis. Brain Topography, 1989, 2: 57-61.

Hjorth, B. The physical significance of time domain descriptors in EEG analysis. Electroenceph. Clin. Neurophysiol., 1973, 34: 321-325.

Hjorth, B. An on-line transformation of EEG scalp potentials into orthogonal source derivations. Electroenceph. Clin. Neurophysiol., 1975, 39: 526-530.

Hjorth, B. Source derivation dimplifies topographical EEG interpretation, Am. J. EEG Technol., 1980, 20: 121-132.

Hjorth, B. An adaptive EEG derivation technique, Electroenceph. Clin. Neurophysiol., 1982, 54: 654-661.

Hjorth, B. and Rodin, E. Extraction of "deep" components for scalp EEG. Brain Topography, 1988, 1: 65-69.

Hjorth, B. and Rodin, E. An Eigenfunction approach to the inverse problem of EEG. Brain Topography, 1989, 1: 79-86.

Hooshmand, H., Beckner, E. and Radfar, F. Technical and clinical aspects of topographic brain mapping. Clin. Electroencephal., 1989, 20: 235-247.

Hughes, J.R. and Miller, J.K. An example of the possible clinical usefulness of topographic EEG displays. Clin. Electroenceph. 1989, 20: 39-44.

IEEE Digital Signal Processing Committee (Ed.), Programs for digital signal processing, 1979.

Ihl, R., Maurer, K., Dierks, T., Frlich, L. and Perisic, I. Staging in dementia of the Alzheimer type - topography of electrical brain activity reflects the severity of the disease. Psychiatry Research, 1989, 29: 399-401.

Imig, T. J. and Morel, A. Organization of the thalamocortical auditory system in the cat. Ann. Rev. Neurosci., 1983, 6: 95-120.

Inouye, T., Shinosaki, Yagasaki, A. and Shimizu, A. Spatial distribution of generators of alpha activity. Electroenceph. Clin. Neurophysiol., 1986, 63: 353-360.

Itil, T.M. and Itil, K.Z. The significance of pharmacodynamic measurements in the assessment of bioavailability and bioequivalence of psychotropic drugs using CEEG and dynamic brain mapping. J. Clin. Psychiat., Supp. 44, 1986.

Itil, T.M., Saletu, B. and Davis, S. EEG findings in chronic schizophrenics based on digital computer period analysis and analog power spectra. Biol. Psychiat., 1972, 5: 1-13.

Iwayama, K., Mori, K., Iwamoto, K., Yamauchi, T. and Masago, M. Origin of frontal N15 component of somatosensory evoked potential in man. Electroenceph. Clin. Neurophysiol., 1988, 71: 125-132.

Jacobson, G.P. and Grayson, A.S. The normal scalp topography of Middle Latency Auditory Evoked Potential Pa component following monaural click stimulation. Brain Topography, 1988, 1: 29-36.

Jayant, N.S. and Noll, P. Digital coding of waveforms. Prentice-Hall, New Jersey, 1984.

Jeffreys, D.A. Polarity and distribution of human visual evoked potential (VEP) components as clues to cortical topography. Electroenceph. Clin. Neurophysiol., 1970, 29: 328.

Jerrett, S.A. and Corsak, J. Clinical utility of topographic EEG brain mapping. Clin. Electroencephal. 1988, 19: 134-143.

Johnson, B.W., Weinberg, H., Ribary, U., Cheyne, D.O. and Ancill, R. Topographic distribution of the 40 Hz auditory evoked-related potential in normal and aged subjects. Brain Topography, 1989, 1: 117-122.

Jones, S.J. and Power, C.N. Scalp topography of human somatosensory evoked potentials: the effect of interfering tactile stimulation applied to the hand. Electroenceph. Clin. Neurophysiol., 1984, 58: 25-36.

Jordan, S.E., Nowacki, R. and Nuwer, M. Computerized electroencephalography in the evaluation of early dementia. Brain Topography, 1989, 1: 271-282.

Joseph, A.B. Mapping neuronal electrical function [letter]. J. R. Soc. Med., 1981, 74: 632.

Joseph, J.P., Remond, A., Rieger, H. and Lesevre, N. The alpha average. II. Quantitative study and the proposition of a theoretical model. Electroenceph. Clin. Neurophysiol., 1969, 26: 350-360.

Kadoya, C., Wada, S. and Matsuoka, S. Clinico-experimental studies on auditory evoked middle latency response (AEMLR) with specific reference to generation and auditory dominancy. Sangyo Ika Daigaku Zasshi, 1988, 10: 11-30.

Kahn, E.M., Weiner, R.D., Brenner, R.P. and Coppola, R. Topographic maps of brain electrical activity - pitfalls and precautions. Biol. Psychiatry, 1988, 23: 628-36.

Kakigi, R. and Shibasaki, H. Scalp topography of the short-latency somatosensory evoked potentials following posterior tibial nerve stimulation in man. Electroenceph. Clin. Neurophysiol., 1983, 56: 430-437.

Kakigi, R. and Shibasaki, H. Scalp topography of mechanically and electrically evoked somatosensory potentials in man. Electroencephal. Clin. Neurophysiol, 1984, 59: 44-56.

Karniski, W. and Clifford Blair, R., Topographic and temporal stability of the P300. Electroenceph. Clin. Neurophysiol., 1989, 72: 373-383.

Karson, C.N., Coppola, R., Morihisa, J.M. and Weinberger, D.R. Computed electroencephalographic activity mapping in schizophrenia, Arch. Gen. Psychiatry, 1987, 44: 514-517.

Katayama, Y., Tsubokawa, T., Maejima, S., Hirayama, T. and Yamamoto, T. Corticospinal direct response in humans: identification of the motor cortex during intracranial surgery under general anaesthesia. J. Neurol. Neurosurg. Psychiatry, 1988, 51: 50-59.

Kaufman, L., Okada, Y., Brenner, D. and Williamson, S. J. On the relation between somatic evoked potentials and fields. Int. J. Neurosci., 1981, 15: 223-239.

Kavanagh, R.N., Darcey, T.M., Lehmann, D. and Fender, D.H. Evaluation of methods for three-dimensional localization of electrical sources in the human brain. IEEET, BME-25, 1978, 421-429.

Kendall M., Stuart A. "The advanced theory of statistics". McMillan Publishing Co., New York, N.Y., 1977, 1: 4th Edition.

Kendall M., Stuart A., Ord J.K. "The advanced theory of statistics" Oxford University Press, New York, N.Y., 1983, 3: 4th Editon.

Kertesz, A. and Lesk, D., Localization of lesions in aphasia. Neurol. Neurocir. Psiquiatr., 1977, 18: 87-96.

Kitani, Y., Watanabe, Y. and Fujita, T. Monitoring of EEG activity during NLA anesthesia recorded on topographic computerized display map. Electroenceph. Clin. Neurophysiol., 1985, 61: 1985.

Knight, R. T. Neurophysiological mechanisms: evidence from human lesion data. Paper presented at the Eighth International Conference on Event-Related Potentials of the Brain, California, U.S., 1986.

Knudsen, E.I., du Lac, S. and Esterly, S.D. Computational maps in the brain. Annu. Rev. Neurosci., 1987, 10: 41-65.

Kohrman, M.H., Sugioka, C., Huttenlocher, P.R. and Spire, J.P. Inter-subject versus intra-subject variance in topographic mapping of the Electroencephalogram. Clin. Electroencephal. 1989, 20: 248-253.

Koles, Z.J. and Paranjape, R.B., Topographic mapping of EEG: An examination of accuracy and precision. Brain Topography, 1988, 1: 87-96.

Kooi, K. and Yamada, T. Field studies of monocularly evoked cerebral potentials in bitemporal hemianopsia. Neuro., 1973, 23: 1217-1225.

Kraus, N. and McGee, T. Color imaging of the human middle latency response. Ear and Hearing, 1988, 9: 159-167.

Kraus, N., McGee, T., Stark, C. and Jacobson, S. Multichannel intracranial recording device using a color imaging brain mapping system. Brain Topography, 1988, 1: 61-64.

Kraus, N., Smith, D.I. and McGee, T. Midline and temporal lobe MLRs in the guinea pig originate from different generator systems: a conceptual framework for new and existing data. Electroenceph. Clin. Neurophysiol., 1988, 70: 541-558.

Krishnaiah P.R., Kanal L.N. (Eds.), "Classification, pattern recognition and reduction of dimensionality". North Holland Publishing Company, Amsterdam, 1982.

Kumar, A. and Dobben, G.D. Central auditory and vestibular pathology. Otolaryngol. Clin. North Am., 1988, 21: 377-389.

Kurtzberg, D. and Vaughan, H.H. Topographic analysis of human cortical potentials preceding self-initiated and visually triggered saccades. Brain Res., 1982, 243: 1-9.

Lai, C.W. The effect of eye/hand dominance on topographic distribution of visual evoked potentials. Electroenceph. Clin. Neurophysiol., 1986, 64: 82.

Lang, W., Lang, M., Uhl, F., Koska, Ch. and Kornhuber, A. Negative cortical DC shifts preceding and accompanying simultaneous and sequential finger movements. Exp. Brain Res. 1988, 71: 579-587.

Languis, M. and Wittrock, M. Integrating neuropsychological and cognitive research: a perspective for bridging brain-behavior relationships. In: J.E. Obrzut and G.W. Hynd (Eds.), Child Neuropsychology: Theory and Research, Academic Press, New York, 1986, 209-239.

Larsen R.J. and Marx M.L. "An introduction to mathematical statistics and its applications". Prentice-Hall, Englewood Cliffs, N.J., 1981.

Lechner, H., Niederkorn, K., Logar, C. and Schimdt, R. Topographic EEG brain mapping in cerebrovascular disease and dementia. Neurologija, 1989, 38: 3-10.

Lee, S. and Buchsbaum, M.S. Topographic mapping of EEG artifacts. Clin. EEG, 1987, 18: 61-67.

Lee, Y.S., Lueders, H., Dinner, D.S., Lesser, R.P., Hahn, J. and Klem, G. Recording of auditory evoked potentials in man using chronic subdural electrodes. Brain, 1984, 107: 115-131.

Lehmann, D. and Michel, C. M. Intracerebral dipole sources of EEG FFT power maps. Brain Topography, 1989, 2: 155-164.

Lehmann, D. Multichannel topography of human alpha EEG fields. Electroenceph. Clin. Neurophysiol., 1971, 31: 439-449.

Lehmann, D. Human scalp EEG fields: evoked, alpha, sleep, and spike-wave patterns. In: H. Petsche and M.A.B. Brazier (Eds.), Synchronization of EEG activity in epilepsies. Springer, Wien and New York, 1972, 307-326.

Lehmann, D. EEG assessment of brain activity: spatial aspects; segmentation and imaging. Int. J. Psychophysiol., 1984, 1: 267-276.

Lehmann, D. The EEG as scalp field distribution. In: A. Remond (Ed.), EEG informatics. A didactic review of methods and applications of EEG data processing. Elsevier/North-Holland Biomed. Pres, 1977, 365-384.

Lehmann, D. Spatial analysis of evoked and Spontaneous EEG Potential Fields. In: N. Yamaguchi and K. Fujisawa (Eds.), Recent advances in EEG and EMG data processing. Elsevier/North-Holland Biom Press, 1981, 117-132.

Lehmann, D. EEG assessment of brain activity: spatial aspects, segmentation and imaging. International J. Psychophysiol., 1984: 267-276.

Lehmann, D. Spatial analysis of EEG and evoked potential data. In: E.F. Duffy (Ed.), Topographic mapping of brain electrical activity., Butterworths, 1986, 29-61.

Lehmann, D. Principles of spatial analysis. In: A. Gevins and A. Redmond (Eds.), Handbook of Electroencephalography and Clinical Neurophysiology: Analysis of Brain Electrical and Magnetic Signals. Elsevier, Amsterdam, 1987, 1-46.

Lehmann, D., Spontaneous EEG momentary maps and FFT power maps. In: D. Samson-Dellfus, J.D. Guieu, J. Gotman, and P. Etevenon (Eds.), statistics and topography in quantitative EEG. Elsevier, Amsterdam, 1988, 27-48.

Lehmann, D. Brain electrical mapping of cognitive functions for psychiatry - functional micro-states. Psychiatry Research, 1989a, 29: 385-386.

Lehmann, D. The view of an EEG-EP mapper. Brain Topography, 1989b, 1: 77-78.

Lehmann, D., Kavanagh, R.N. and Fender, D.H. Field studies of averaged visually evoked EEG potentials in a patient with a split chiasm. Electroenceph. Clin. Neurophysiol., 1969, 26: 193-199.

Lehmann, D., Meles, H.P. and Mir, Z. Average multichannel EEG potential fields evoked from upper and lower hemi-retina: latency differences. Electroenceph. Clin. Neurophysiol., 1977, 43: 725-731.

Lehmann, D., Ozaki, H. and Pal, I. Averaging of spectral power and phase via vector diagram best fits without reference electrode or reference channel. Electroenceph. Clin. Neurophysiol., 1986, 64: 350-363.

Lehmann, D., Ozaki, H. and Pal, I. EEG alpha map series: brain micro-states by space-oriented adaptive segmentation. Electroenceph. Clin. Neurophysiol., 1987, 67: 271-288.

Lehmann, D. and Skrandies, W. Multichannel evoked potential fields show different properties of human upper and lower hemiretina systems. Exp. Brain Res., 1979, 35: 151-159.

Lehmann, D. and Skrandies, W. Reference-free identification of components of checkerboard-evoked multichannel potential fields. Electroenceph. Clin. Neurophysiol., 1980, 48: 609-621.

Lehmann, D. and Skrandies, W. Spatial analysis of evoked potentials in man - a review. Prog. Neurobiol., 1984, 23: 227-250.

Lehmann E.I. "Theory of point estimation" John Wiley and Sons, New York, N.Y., 1983.

Lehmann E.I. "Testing statistical hypotheses". John Wiley and Sons, New York, N.Y., 1959.

Lehmann E.I. "Nonparametrics: statistical methods based on ranks". Holden-Day, San Francisco, CA., 1975.

Lemieux, J.F. and Blume, W.T. Topographical evolution of spike-wave complexes. Brain Res., 1986, 373: 1-2.

Lemieux, J.F., Vera, R.S. and Blume, W.T. Technique to display topographic evolution of EEG events. Electroenceph. Clin. Neurophysiol., 1984, 58: 565-568.

Levy, J. Brain asymmetries and fallacies in statistical inference, Brain Cogn., 1986, 5: 115-25.

Lilly, J.C. Channel recorder for mapping the electric potential gradients of the cerebral cortex: electro-iconograms. EE, 1950, 69: 68-69.

Lilly, J.C. A method of recording the moving electrical potential gradients in the brain: the 25 channel bavatron and electro-iconograms. In: Conference on Electronics in Nucleonics and Medicine. Amer. Inst. of Electr. Eng., New York, 1950: 37-43.

Livanov, M.N. Ananiev, B.M. and Bechtereva, N.P. Study of the bioelectrical mosaic of the cortex in patients with brain tumours and traumata with the aid of encephaloscopy. Sn. Neuropat. Psikhiat., 1956, 56: 778-809.

Lombroso, C.T. Sylvin seizures and midtemporal spikes foci in children. Arch. Neurol. 1967, 17: 52-59.

Lombroso, C.T. and Duffy, F.H. Brain electrical activity mapping in the epilepsies. Advances in epileptology: XIIIth Epilepsy International Symposium Raven Press, New York, 1982.

Lombroso, C.T. and Duffy, F.H. Brain electrical activity mapping as an adjunct to CT scanning. In: R. Canger, F. Angeleri and J.K. Penry (Eds.), Advances in Epileptology: XIIth Epilepsy International Symposium. Raven Press, New York, 1980, 83-88.

Lopes da Silva, F. Computerized EEG analysis (CEAN) in epilepsy. 17th Epilepsy International Congress, Jerusalem, Israel, September 6-11th, 1987.

Lopes da Silva, F., Pijin, J.P. and Boeijinga, P. Interdependence of EEG signals: linear v. nonlinear associations and the significance of time delays and phase shifts. Brain Topography. 1989, 2: 9-18.

Lounasmaa, O.V., Williamson, S.J., Kaufman, L. and Tanenbaum, R. Visually evoked responses from non-occipital areas of thhuman cortex. In: H. Weinberg, G. Stroink and T. Katila (Eds.), Biomagnetism: Applications and Theory, Pergamon Press, New York, 1985, 348-353.

Lovrich, D., Novick, B. and Vaughan, H.G., Jr. Topographic analysis of auditory event-related potentials associated with acoustic and semantic processing. Electroenceph. Clin. Neurophysiol., 1988, 71: 40-54.

Luders, H., Daube, J.R., Taylor, W.F. and D.W. Klass. A computer system for statistical analysis of EEG transients. In: P. Kellaway and Ingemar Petersen (Eds.), Quantitative Analytic Studies in Epilepsy. Raven Press, New York, 1976: 403-429.

Luders, H., Lesser, R.P., Dinner, D.S., Morris, H.H., Wyllie, E. and Godoy, J. Localization of cortical function: new information from extraoperative monitoring of patients with epilepsy. Epilepsia, 1988, Supp. 29: S56-65.

Lukas, S.E., Mendelson, J.H., Woods, B.T., Mello, N.K. and Teoh, S.K. Topographic distribution of EEG alpha activity during ethanol-induced intoxication in women. J. Stud. Alcohol., 1989, 50: 176-185.

Luria, A.R. The localization of function in the brain. Biol. Psychiatry, 1978, 13: 633-635.

MacKay, D.M. On-line source-density computation with a minimum of electrodes. Electroenceph. Clin. Neurophysiol., 1983, 56: 696-698.

MacKay, D.M. Source density analysis of scalp potentials during evaluated action. II. Lateral distributions. Exp. Brain Res., 1984, 54: 86-94.

Malloy, P., Rasmussen, S., Braden, W. and Haier, R.J. Topographic evoked-potential mapping in obsessive - compulsive disorder - evidence of frontal-lobe dysfunction. Psychiatry Research, 1989, 28: 63-71.

Manmaru, S. and Matsuura, M. Quantification of benzodiazepine-induced topographic EEG changes by a computerized wave form recognition method - application of a principal component analysis. Electroenceph. Clin. Neurophysiol., 1989, 72: 126-132.

Mars, N.J.I. and Lopes da Silva, F.H. Propagation of seizure activity in kindled dogs. Electroenceph. Clin. Neurophysiol., 1983, 56: 194-209.

Mars, N.J.I. and Lopes da Silva, F.H. EEG analysis methods based on information theory. In: A.S. Gevins and A. Remond (Eds.), Methods of analysis of brain electrical and magnetic signals. EEG Handbook, Elsevier Science Publisher, Amsterdam, 1987, (revised series) 1: 297-307.

Mateer, C.A. Functional organization of the right nondominant cortex: evidence from electrical stimulation. Can. J. Psychol., 1983, 37: 36-58.

Matsuoka, S., Arakaki, Y., Numaguchi, K. and Ueno, S. The effect of dexamethasone on electroencephalograms in patients with brain tumors, J. Neurosurg., 1978, 48: 601-608.

Matthews, P.B., Proprioceptors and their contribution to somatosensory mapping: complex messages require complex processing., Can. J. Physiol. Pharmacol., 1988, 66: 430-438.

Maurer, K. Topography of EEG and evoked potentials in dementia before and under treatment with encephalotropics. In: K. Maurer and R.J. Wurtman (Eds.), Organic Brain Disorders, Vieweg, Braunschweig, 1989, 27-36.

Maurer, K. and Dierks, T. Brain mapping - topographic demonstration of the EEG and evoked potentials in psychiatry and neurology, EEG EMG. 1987a, 18: 4-12.

Maurer, K. and Dierks, T. Functional imaging of the brain in psychiatry - mapping of EEG and evoked-potentials. Neurosurgical Review, 1987b, 10: 275-282.

Maurer, K., Dierks, T. and Ihl, R. Quantitative P300 data and their topography in dementia. In: D. Samson-Dollfus, J. D. Guieu, J. Gotman, and P. Etevenon (Eds.), Statistics and Topography in Quantitative EEG. Elsevier, Amsterdam, 1988, 243-250.

McCallum, W.C. Potentials related to expectancy, preparation and motor activity. In: T.W. Picton (Eds.), Human Event-Related Potentials. 427-517.

McCarley, R.W., Winkelman, J.W. and Duffy, F.H. Human cerebral potentials associated with REM sleep rapid eye movements: links to PGO waves and waking potentials, BR, 1983, 274: 359-364.

McClelland, G.R., Raptopoulos, P. and Jackson, D. The effect of paroxetine on the quantitative EEG. Acta Psychiatrica Scandinavica, 1989, 80: 50-52.

McCullagh, P.J. and McClelland, R.J. Topographical brain electrical activity mapping on an IBM-compatible personal computer. J. Biomed. Eng., 1989, 11: 137-40.

McGee Jr., F.E., Lee, R.G., Harris, J.A., Melby, G. and Bickford, R.G. Applications of interpolation and read display in EEG. Electroenceph. Clin. Neurophysiol., 1969, 27: 544.

Meijs, J.W.H., Bosch, F.G.C., Peters, M.J. and Lopes da Silva, F. On the magnetic field distribution generated by a dipolar current source situated in a realistically shaped compartment model of the head. Electroenceph. Clin. Neurophysiol., 1987, 66 (3): 286-298.

Meles, P.H. and Wieser, H.G. Computer-generated dynamic presentation of functional versus anatomical distances in the human brain. Appl. Neurophysiol., 1982, 45: 404-405.

Meyer, A. The frontal lobe syndrome, the aphasias and related conditions. A contribution to the history of cortical localization. Brain, 1974, 97: 565-600.

Miller, L. Narrow localizationism' in psychiatric neuropsychology. Psychol Med, 1986, 16: 729-34.

Mimbrera, G.B., Deleon, M.G., Gutierrez, S.D., Vazquez, J. M., Salcines, J.T. and Fernandez, J.M.D. Evaluation of the therapeutic effect of flumazenil in hepatic encephalopathy by brain mapping (in Spanish), Revista Espanola de las Enfermedades del Aparato Digestivo, 1989, 75: 277-279.

Modena, I., Ricci, G. B., Barbanera, S., Leoni, R. and Carelli, P. Biomagnetic measurements of spontaneous brain activity in epileptic patients. Electroenceph. Clin. Neurophysiol., 1982, 54: 622-628.

Montgomery D.C. "Design and analysis of experiments" 2nd edition, John Wiley and Sons, New York, N.Y., 1984.

Mood A.M., Graybill F.A. and Boes D.C. "Introduction to the theory of statistics". 3rd Edition, McGraw-Hill, New York, N.Y., 1974.

Morihisa, J.M., Duffy, F.H., Mendelson, W.B. and Wyatt, R. J. The use of brain electrical activity mapping (BEAM) as an exploratory technique to delineate regional differenced between schizophrenic patients and control subjects. In: P. Flor-Henry and J. Gruzelier (Eds.), Laterality and Psychopathology. Elsevier, New York, 1983.

Morihisa, J.M., Duffy, F.H. and Wyatt, R.J. Topographic analysis of computer processed electroencephalography in schizophrenia. In: E. Usdin and J. Hanin (Eds.), Biological Markers in Psychiatry and Neurology, Pergamon Press, New York, 1982.

Morihisa, J.M., Duffy, F.H. and Wyatt, R.J. Brain electrical activity mapping (BEAM) in schizophrenic patients. Arch. Gen. Psych., 1983, 40 (7): 719-728.

Morihisa, J.M., Duffy, F.H. and Wyatt, R.J. Brain electrical activity mapping in psychiatry. In: J.M. Morihisa (Ed.), Brain Imaging in Psychiatry. American Psychiatric Press, Washington, 1984, 78-93.

Morihisa, J.M. and McAnulty, G.B. Structure and function: Brain electrical activity mapping and computed tomography in schizophrenia, Biol. Psychiat., 20/1, 1985.

Morrison D.F. "Multivariate Statistical methods". 2nd Edition, McGraw-Hill, New York, N.Y., 1976.

Morstyn, R. Altered topography of EEG spectral content in schizophrenia. Electroenceph. Clin. Neurophysiol., 1983, 56: 263-271.

Morstyn, R., Duffy, F.H. and McCarley, R.W. Altered P300 topography in schizophrenia. Arch. Gen. Psych., 1983a, 40: 729-734.

Morstyn, R., Duffy, F.H. and McCarley, R.W. Altered topography of EEG spectral content in schizophrenia. Electroenceph. Clin. Neurophysiol., 1983b, 56: 263-271.

Mueller, T., Dierks, T., Fritze, J. and Maurer, K. Functional brain imaging (Mapping of EEG) in relation to psychopathologic changes in schizophrenia. Psychiatry Research, 1989, 29: 419-420.

Munck, J.C. A mathematical and physical interpretation of the electromagnetic field of the brain. J.T.G. Horikx, 1989.

Myers R.H. "Classical and modern regression with applications" Duxbury Press, Boston, Mass., 1986.

Myslobodsky, M.S., Coppola, R., Bar-Ziv, J., Karson, C., Daniel, D., van Praag, H. and Weinberger, D.R. EEG asymmetries may be affected by cranial and brain parenchymal asymmetries. Brain Topography, 1989, 1: 221-228.

Myslobodsky, M.S., Bar-Ziv, J., van Praag, H. and Glicksohn, J. Bilateral Alpha Distribution and Anatomic Brain Asymmetries. Brain Topography, 1989, 1: 229-235.

Nadvornik, P., Sramka, M. and Fritz, G. Graphic representation of the epileptic focus. Acta Neurochir (Wien), 1976, 21-25.

Nagata, K. Topographic EEG in brain ischemia - correlation with blood flow and metabolism. Brain Topography, 1989, 1: 97-106.

Nagata, K. Topographic EEG mapping in cerebrovascular disease. Brain Topography, 1989, 2: 119-128.

Nagata, K., Gross, C.E., Kindt, G.W., Geier, J.M. and Adey, G.R. Topographic electroencephalographic study with power ratio index mapping in patients with malignant brain tumors. Neurosurg., 1985, 17: 613-619.

Nagata, K., Mizukami, M., Araki, G., Kawase, T. and Hirano, M. Topographic encephalographic study of cerebral infarction using computed mapping of the EEG (CME), J. Cerebral Blood Flow Metab., 1982, 2: 79-88.

Nagata, K., Tagawa, K., Hiroi, S., Shishido, F. and Uemura, K. Electroencephalographic correlates of blood-flow and oxygen-metabolism provided by positron emission tomography in patients with cerebral infarction. Electroenceph. Clin. Neurophysiol., 1989, 72: 16-30.

Nagata, K., Yunoki, K., Araki, G. and Mizukami, M. Topographic electroencephalographic study of transient ischemic attacks, Electroenceph. Clin. Neurophysiol., 1984, 58: 291-301.

Nakano, S., Okuno, T. and Mikawa, H. Landau-Kleffner Syndrome EEG topographic studies. Brain and Development, 1989, 11: 43-50.

Narici, L. and Romani, G.L. Neuromagnetic investigation of synchronized spontaneous activity. Brain Topography, 1989 2: 19-30.

Naylor, D.E., Lieb, J.P. and Risinger, M. Computer enhancement of scalp-sphenoidal ictal EEG in patients with complex partial seizures, 1988, 70: 205-219.

Neill, R.A. and Fenelon, B. Scalp response topography to dynamic random dot stereograms. Electroenceph. Clin. Neurophysiol, 1988, 69: 209-217.

Neshige, R. Lueders, H. and Shibasaki, H. Recording of movement-related potentials from scalp and cortex in man. Brain. 1988, 111: 719-736.

Nunez, P.L. Estimation of large scale neocortical source activity with EEG surface laplacians. Brain Topography, 2: 141-154.

Nunez, P.L. Electric fields of the brain. Oxford University Press, New York, 1981.

Nunez, P.L. Methods to estimate spatial properties of dynamic cortical generator activity. In: G. Pfurtscheller and F.H. Lopes da Silva (Eds.), Functional Brain Imaging, 3-10. Hans Huber Publishers, Toronto, 1988.

Nunez, P.L. Generation of human EEG by a combination of long and short range neocortical interactions. Brain Topography, 1: 1989, 199-215.

Nuwer, M.R. A Comparison of the analyses of EEG and evoked potentials using colored bars in place of colored heads. Electroenceph. Clin. Neurophysiol., 1985, 61: 310-313.

Nuwer, M.R. Frequency analysis and topographic mapping of EEG and evoked potentials in epilepsy. Electroenceph. Clin. Neurophysiol., 1988a, 69: 118-126.

Nuwer, M.R. Quantitative EEG. 1. Techniques and problems of frequency analysis and topographic mapping. J. Clin. Neurophysiol., 1988b, 5: 1-43.

Nuwer, M.R. Quantitative EEG. 2. Frequency analysis and topographic mapping in clinical settings. J. Clin. Neurophysiol., 1988c, 5: 45-85.

Nuwer, M.R. Uses and abuses of brain mapping. Archives of Neurology, 1989, 46: 1134-1136.

Nuwer, M.R. and Jordan, S.E. The centrifugal effect and other spatial artifacts of topographic EEG mapping. J. Clin. Neurophysiol., 1987, 4: 321-326.

Nuwer, M.R., Jordan, S.E. and Ahn, S.S. Evaluation of stroke using EEG frequency analysis and topographic mapping. Neurology, 1987, 37: 1153-1159.

Nuwer, P.L. Electric fields of the brain. New York: Oxford Univserity Press, 1981.

Offner, F.F. The EEG as potential mapping: The value of the average monopolar reference. Electroenceph. Clin. Neurophys, 1959, 2: 213-217.

Ogura, H. and Hatta, T. A new test for hemisphericity: the examination of its validity and reliability. Shinrigaku Kenkyu, 1983, 54: 36-42.

Ojemann, G.A. Mapping of neuropsychological language parameters at surgery. Int. Anesthesiol. Clin., 1986, 24: 115-131.

Ojemann, G.A. and Whitaker, H.A. Language localization and variability. Brain Lang, 1978, 6: 239-60.

Okada, Y.C. Discrimination of localized and distributed current dipole generators and localized single and multiple generators. In: H. Weinberg, G. Stroink and T. Katila (Eds.), Biomagnetism: Applications and Theory, Pergamon Press, New York, 1985, 266-272.

Okada, Y.C., Kaufman, L., Brenner, D. and Williamson, S. J. Application of a SQUID to measurement of somatically evoked fields: transient responses to electrical stimulation of the median nerve: In: S.N. Erne', H.D. Hahlmohm and H. Lubbig (Eds.), Biomagnetism, Walter de Gruyter, Berlin, 1981, 445-461.

Okada, Y., Kaufman, L., Brenner, D. and Williamson, S. J. Modulation transfer functions of the human visual system revealed by magnetic field measurements. Vision Res., 1982 22: 319-333.

Okada, Y. C., Kaufman, L. and Williamson, S. J. The hippocampal formation as a generator of the slow endogenous potentials. Electroenceph. Clin. Neurophysiol., 1983, 55: 417-426.

Okada, Y., Williamson, S. J. and Kaufman, L. Magnetic field of the human sensorimotor cortex. Intl. J. Neurosciences, 1982, 17: 33-38.

Oppenheim, A.V., Schafer, R.W. Digital signal processing. Prentice-Hall, New Jersey, 1975.

Orman, S. S. and Humphrey, G. L. Effects of changes in cortical arousal and of auditory cortex cooling on neuronal activity in the medial geniculate body. Exp. Brain Res., 1981, 42: 475-482.

Ormejohnson, D.W. and Gelderloos, P. Topographic EEG   Brain Mapping During Yogic Flying. International J. Neurosci., 1988, 38: 427-434.

Pantev, C., Hoke, M., Lehnertz, K., Lutkenhoner, B.,   Anogianakis, G. and Wittkowski, W. Tonotopic organization   of the human auditory cortex revealed by transient auditory evoked magnetic fields. Electroenceph. Clin. Neurophysiol., 1988, 69: 160-170.

Paranjape, R.B. and Koles, Z.J. Topographic EEG mapping using the 10-20 system.   Western EEG Society Annual Meeting, Palm Desert, California. February 26-28, 1986.

Pascualmarqui, R.D., Gonzalezandino, S.L., Valdessosa, P.A. and Biscaylirio, R. Current source density-estimation and interpolation based on the spherical harmonic fourier expansion. International J. Neurosci., 1988, 43: 237-249.

Paternoster, R.H., Goossens, M.J., Steenhaut, O.L. and Van   Oost, E.J. A 21-channel EEG-monitor with real-time result   colour display.   In: J. Tiberghien, G. Carlstedt, and J. Lewi (Eds.), Microprocessors and their Applications, North-Holland, 1979, 169-177.

Patterson, T., Spohn, H.E. and Hayes, K., Topographic   evoked potentials during backward masking in schizophrenics, patient controls and normal controls. Prog. Neuro-Psychopharmacol. Biol. Psychiat., 1987, 11: 709-728.

Pernier, J., Perrin, F. and Bertrand, O. Scalp current density fields: concept and properties. Electroenceph. Clin. Neurophysiol., 1988, 69: 385-389.

Perrin, F., Pernier, J., Bertrand, O., Giard, M.H. and Echallier, J.F. Mapping of scalp potentials by surface spline interpolation. Electroenceph. Clin. Neurophysiol., 1987, 66: 75-81.

Persson, A. and Hjorth, B. EEG topogram - an aid in   describing EEG to the clinician. Electroenceph. Clin. Neurophysiol., 1983, 56: 399-405.

Petsche, H. Topography of the EEG: Survey and Prospects, Electroenceph. Clin. Neurophysiol., 1976, 76: 15-28.

Petsche, H. and Marko, A. Das Photozellentoposcop, eine einfache Methode zur Bestimmung der Feldverteilung and Ausbreitung hirnelektrischer vorgange. Arch. Psychiat. Nervenkr., 1954, 192: 447-462.

Petsche, H., Rappelsberger, P. and Trappl, R. Properties of cortical seizure potential fields. Electroenceph. Clin. Neurophysiol., 1970, 29: 567-578.

Petsche, H. and Stumpf, C. Topographic and toposcopic study of origin and spread of the regular synchronized arousal pattern in the rabbit. Electroenceph. Clin. Neurophysiol., 1960, 12: 589-600.

Pfurtscheller, G. Spatiotemporal analysis of alpha frequency components with the ERD technique. Brain Topography. 1989 2: 3-8.

Pfurtscheller, G. Mapping of event-related desynchronization and type of derivation. Electroenceph. Clin. Neurophysiol., 1988, 70: 190-193.

Pfurtscheller, G. Functional topography during sensorimotor activation studied with event-related desynchronization mapping. J. Clin. Neurophysiol., 1989, 6: 75-84.

Pfurtscheller, G. and Aranibar, A. Event-related cortical desynchronization detected by power measurements of scalp EEG. Electroenceph. Clin. Neurophysiol., 1977, 42: 817-826.

Pfurtscheller, G. and Aranibar, A. Evaluation of event-related desynchronization (ERD) preceding and following voluntary self-paced movement. Electroenceph. Clin. Neurophysiol., 1979, 46: 138-146.

Pfurtscheller, G. and Cooper, R. Frequency dependence of the transmission of the EEG from cortex to scalp. Electroenceph. Clin. Neurophysiol., 1975, 93-96.

Pfurtschller, G., Steffan, J. and Maresch, H. ERD-mapping and functional topography - temporal and spatial aspects. In: G. Pfurtschller and F.H. Lopes da Silva (Eds.), Functional Brain Imaging. Huber, Toronto, 1988b: 117-130.

Pijn, J.P.M., Vijn, P.C.M., Lopes da Silva, F.H., Van Emde Boas, W. and Blanes, W. The use of signal-analysis for localization of an epileptogenic focus: a new approach. Advances in epileptology, 1989, 17: 272-276.

Pike, M.G., Wong, P.K.H., Bencivenga, R., Flodmark, O., Cabral, D.A., Speert, D.A. and Farrell, K. Electrophysiologic studies computed tomography and neurologic outcome in acute bacterial meningitis. J. Pediatrics. (In press, 1990).

Pockberger, H. Petsche, H., Rappelsberger, P. Intracortical aspects of penicillin-induced seizure patterns in the rabbit's motor cortex. In: E.J. Speckman and C.E. Elger (Eds.), Epilepsy and Motor System. Urban and Schwarzenberg Publishers, Germany, 1983: 161-178.

Pockberger, H., Petsche, H., Rappelsberger, P., Zidek, B. and Zapotoczky, H.G. On-going EEG in depression: a topographic spectral analytical piolt study. Electroenceph. Clin. Neurophysiol., 1985, 61: 349-358.

Pockberger, H., Rappelsberger, P., Petsche, H., Thau, K. and Kufferle, B.

Computer-assisted EEG topography as a tool in the evaluation of actions of psychoactive drugs in patients. Neuropsychobiol., 1984, 12: 183-187.

Podreka, I., Lang, W., Suess, E., Wimberger, D., Steiner, M., Gradner, W., Zeitlhofer, J., Pelzl, G., Mamoli, B. and Deecke, L. Hexa-methyl-propylene-amine-oxime (HMPAO) single photon emission computed tomography (SPECT) in epilepsy. Brain Topography, 1: 1988, 55-60.

Poliac, M., Popoviciu, L., Corfariu, O. and Tudosie, M. Electro-encephalographic map - a new computer processing method for diagnosis in clinical and experimental neurophysiology. Neurol. Psychiatr., (Bucur), 1981, 19: 289-302.

Pool, K.D., Finitzo, T., Chi-Tzong-Hong, Rogers, J. and Picket, R.B. Infarction of the superior temporal gyrus: a description of auditory evoked potential latency and amplitude topology. Ear and Hearing, 1989, 10: 144-152.

Pool, K.D., Aronofsky, J.S., Finitzo, T. and Barr, R.S. Localization of multiple dipoles: mathematical programming approaches. Brain Topography, 1989, 1: 247-255.

Popoviciu, L., Tudosie, M., Schiopu, M. and et al. The contribution of the electroencephalographic computerized map in setting down the topography of the cerebral infarctions. Rev. Roum. Med., 22/2, 1984.

Porjesz, B. and Begleiter, H. Alcohol and bilateral evoked brain potentials. Adv. Exp. Med. Biol. 1975, 59: 553-567.

Pritchard, W.S., P300 and EPQ/STPI personality traits. Person. Individ. Diff., 1989, 10: 15-24.

Rabiner, L.R., Gold, B. Theory and practice of digital signal processing, Prentice-Hall, New Jersey, 1975.

Ragot, R.A. and Remond, A. EEG field mapping. Electroenceph. Clin. Neurophysiol., 1978, 45: 417-421.

Rapp, P.E., Bashore, T.R., Martinerie, J.M., Albano, A.M., Zimmerman, I.D. and Mees, A.I. Dynamics of brain electrical activity. Brain Topography, 1989, 2: 99-118.

Rappelsberger, P. and Petsche, H. Probability mapping: power and coherence analyses of cognitive processes.

Rappelsberger, P., Pockberger, H. and Petsche, H. Intracranial recording and generators of the EEG. Electroenceph. Clin. Neurophysiol., 1987, 39: 149-155.

Rappelsberger, P. The reference problem and mapping of coherence: a simulation study. Brain Topography, 1989 2: 63-72.

Reite, M., Edrich, J., Zimmerman, J. T. and Zimmerman, J. E.   Human magnetic auditory evoked fields. Electroenceph. Clin. Neurophysiol., 1978, 45(20): 114-117.

Reite, M, Zimmerman, J. E., Edrich, J. and Zimmerman, J. E.   The human magnetoencephalograph: Some   EEG   and   related   correlations. Electroenceph. Clin. Neurophysiol., 1976, 40: 59-66.

Remond, A. Discussion sur la technique electroencephalographique.   Rev. Neurol., 1951, 84: 580-584.

Remond, A. Integrated and topological analysis of the EEG.   Electroenceph. Clin. Neurophysiol., 1961, 20: 64-67.

Remond, A. Surface electrical fields as displayed by digital computer pergamon. Oxford, 1965.

Remond, A. The importance of topographic data in EEG phenomena, and an electrical model to reproduce them.   Electroenceph. Clin. Neurophysiol., 1969, Supp. 27: 29-49.

Remond, A., Lesevre, N., Joseph, J.P., Rieger, H. and Lairy, G.C. The alpha average. I methodology and description. Electroenceph. Clin. Neurophysiol., 1969, 26: 245-265.

Remond, A. and Offner, F. A new method for EEG display.   Electroenceph. Clin. Neurophysiol., 1952, 7: 453-460.

Remond, A. and Torres, F. A method of electrode placement with a view to topographical   research.   Electroenceph. Clin.   Neurophysiol.,   1964,   17: 577-578.

Ricci, G. B. Clinical magnetoencephalography.   Il Nuovo Cimento, 1983, 2(2): 517-537.

Ricci, G.B., S. Buonomo, S., Peresson, M., Romani, G. L., Salustri, C. and Modena I.   Multichannel neuromagnetic investigation of focal epilepsy. Med. and Biological Computing, 1985, 23 Supp. 1: 42-44.

Ricci, G.B., Romani, G.L., Salustri, C., Pizzella, V., Torrioli, G., Buonomo, S., Peresson, M. and Modena, I. Study of focal epilepsy by multichannel neuromagnetic measurements. Electroenceph. Clin. Neurophysiol., 1987, 66: 358-68.

Rodin, E. and Ancheta, O. Cerebral electrical fields   during petit mal absences. Electroenceph. Clin. Neurophysiol., 1987, 66: 457-466.

Rodin, E. and Cornellier, D. Source derivation recordings of generalized spike-wave complexes. Electroenceph. Clin. Neurophysiol., 1989.

Rodin, E. and Nigro, M. The cerebral electrical fields in cerebellar syndrome. Clinical EEG, 1987, 18: 142-146.

Rodin, E., Wasson, S., Porzak, J., Grisell, J. and Gudobba, R. The cerebral somato-sensory response induced by passive joint movement. Electroenceph. Clin. Neurophysiol., 1970, 28: 91.

Rodin, E.A. Computer assisted clinical neurophysiology - The role of the technologist. J. Electrophysiol. Technol., 1988, 14: 91-108.

Romani, G.L., and Leoni, R. Localization of cerebral courses by neuromagnetic measurements. In: H. Weinberg, G. Stroink and T. Katila (Eds.), Biomagnetism: Applications and Theory, Pergamon Press, New York, 1985, 205-220.

Romani, G.L. and Rossini, P. Neuromagnetic functional localization: principles, state of the art, and perspectives. Brain Topography, 1988, 1: 5-21.

Romani, G.L. and Williamson, S. J. (Eds.), Proceedings of the fourth international Workshop on Biomagnetism. Il Nuovo Cimento, 1983, 2D, 2: 123-664.

Romani, G.L., Williamson, S. J. and Kaufman, L. Tonotopic organization of the human auditory cortex. Science, 1982, 216: 1339-1340.

Romani, G. L., Williamson, S. J., Kaufman, L. and Brenner, D. Characterization of the human auditory cortex by the neuromagnetic method. Exp. Brain Res., 1982, 47: 381-393.

Rose, D.E., Ducla-Soares, E. and Sato, S. Improved accuracy of MEG localization in the temporal region with inclusion of volume current effects. Brain Topography, 1989, 1: 175-181.

Rossini, P.M., Babiloni, F., Bernardi, G., Cecchi, L., Johnson, P.B., Malentacca A., Stanzione, P. and Urbano, A. Abnormalities of short-latency somatosensory evoked potentials in parkinsonian patients. Electroenceph. Clin. Neurophysiol., Jul-Aug 1, 74: 277-89.

Rumsey, J.M., Coppola, R., Denckla, M.B., Hamburger, S.D. and Kruesi, M.J.P. EEG spectra in severely dyslexic men - rest and work and design recognition. Electroenceph. Clin. Neurophysiol., 1989, 73: 30-40.

Ryugo, D. K. and Weinberger, N. M. Corticofugal modulation of the medial geniculate body. Exp. Neurol., 1976, 51: 377-391.

Saletu, B., Anderer, P. and Gruenberger, J., EEG Brain mapping in gerontopsychopharmacology - on protective properties of pyritinol against hypoxic hypoxidosis. Psychiatry Research, 1989, 29: 387-390.

Saletu, B., Anderer, P., Kinsperger, K. and Grunberger, J. Topographic brain mapping of EEG in neuropsychopharmacology, Part II. Clinical applications (pharmaco EEG imaging). Methods Find Exp. Clin. Pharmacol., 1987, 9: 385-408.

Saletu, B., Anderer, P., Kinsperger, K., Grunberger, J. and Sieghart, W. Comparative bioaailability studies with a new mixed-micelles solution of Diazepam utilizing radioreceptor assay, psychometry and EEG brain mapping. Internl. Clin. Psychopharmacol., 1988, 3: 287-323.

Saletu, B., Brunberger, J., Anderer, P. and Barbanoj, M.J. Pharmacodynamic studies of a combination of Lorazepam and Diphenhydramine, current therapeutic research - clinical and experimental. 1988, 44: 909-937.

Saletu, B., Darragh, A., Salmon, P. and Coen, R. EEG Brain mapping in evaluating the time-course of the central action of DUP-996 - a new acetylcholine releasing drug. Br. J. Clin. Parmacol., 1989, 28: 1-16.

Saletu, B. and Grunberger, J. Drug profiling by computed electroencephalography and brain maps, with special consideration of Sertraline and its psychometric effects. J. Clin. Psychiatry, 1988, 49: 59-71.

Saletu, B., Grunberger, J., Anderer, P. and Barbanoj, M.J. Pharmacodynamic studies of a combination of Lorazepam and Diphenhydramine and its single components - electroencephalographic brain mapping and safety evaluation. Current Therapeutic Research-Clinical and Experimental, 1988, 44: 909-937.

Saletu, B., Grunberger, J. and Linzmayer, L. On the central effects of a new partial Benzodiazepine agonist RO 16-6028 in man - pharmaco-EEG and psychometric studies. International J. Clin. Pharmacology Therapy and Toxicology, 1989, 27: 51-65.

Samson-Dollfus, D.A. and Bendoukha, H. Choice of the reference for EEG mapping in the newborn: an initial comparison of common nose reference, average and source derivation. Brain Topography, 1989, 2: 165-169.

Sandini, G., Romano, P., Scotto, A. and Traverso, G. Topography of brain electical activity: a bioengineering approach. Med. Prog. Tech., 1983, 10: 5-19.

Sannita, W.G., Ottonello, D., Perria, B., Rosadini, G. and Timitilli, C. Topographic approaches in human quantitative pharmaco-electro-encephalography. Neuropsychobiol., 1983, 9: 66-72.

Sato, K., Kitajima, H., Mimura, K., Hirota, N., Tagawa, Y. and Ochi, N. Cerebral visual evoked potential in relation to EEG. Electroenceph. Clin. Neurophysiol., 1971, 30: 123-138.

Sato, S. and Smith, P.D. Magnetoencephalography. J. Clin. Neurophysiol., 2/2, 1985.

Scherg, M. and VonCramon, D. A new interpretation of the generators of BAEP waves I-V: results of a spatio-temporal dipole model. Electroenceph. Clin. Neurophysiol., 1985a, 62: 290-299.

Scherg, M. and VonCramon, D. Two bilateral sources of the late AEP as identified by a spatio-temporal dipole model. Electroenceph. Clin. Neurophysiol., 1985b, 62: 32-44.

Schneider, M. Effects of inhomogeneities on surface signals coming from a cerebral dipole source. IEEE Trans. Biomed. Eng., 1974, BME 2: 52-54.

Schwartz, E.L. Spatial mapping in the primate sensory projection: analytic structure and relevance to perception. Biol. Cybern, 1977, 25: 181-94.

Schwartz, E.L., Cortical mapping and perceptual invariance: a reply to Cavanagh. Vision Res, 1983, 23: 831-5.

Sedgewick, R. Algorithms. Addison-Wesley, 1983.

Shaw, J.C., A method for continuously recording characteristics of EEG topography. Electroenceph. Clin. Neurophysiol., 1970, 29: 592-601.

Shaw, J.C. and Roth, M. Potential distribution analysis. Electroenceph. Clin. Neurophysiol., 1955, 7: 285-292.

Shenton, M.E., Faux, S.F., McCarley, R.W., Ballinger, R., Coleman, M. and Duffy, F.H. Clinical correlations of auditory - P200 topography and left temporo - central deficits in schizophrenia - a preliminary-study. J. Psychiatric Research, 1989, 23: 13.

Shenton, M.E., Faux, S.F., McCarley, R.W., Ballinger, R., Coleman, M., Torello, M. and Duffy, F.H. Correlations between abnormal auditory-P300 topography and positive symptoms in schizophrenia - a preliminary-report. Biological Psychiatry, 1989, 25: 710-716.

Shewmon D. and Krentler, K.A. Off-line montage reformatting. Electroenceph. Clin. Neurophys., 1984, 57: 591-595.

Shibasaki, H., Barrett, G., Halliday, E. and Halliday, A. M. Components of the movement-related cortical potential and their scalp topography. Electroenceph. Clin. Neurophysiol., 1980, 49: 213-226.

Shipton, H.W. A new frequency-selective toposcope for electroencephalography. Med. Electron. Biol. Eng., 1963, 1: 483-495.

Shipton, H.W. and Armstrong, G.L. A modern frequency and phase indicating toposcope. Electroenceph. Clin. Neurophysiol., 1981, 52: 659-662.

Sidman, R.D., Giambalvo, V., Allison, T. and Bergey, P. A method for localization of sources of human cerebral potentials evoked by sensory stimuli. Sens Processes, 1978, 2: 116-29.

Silberman, B.W. Density estimation for statistics and data analysis. Chapman and Hall, London, 1986.

Simson, R., Vaughan, H.G. and Ritter, W. The scalp topography of potentials associated with missing visual or auditory stimuli. Electroenceph. Clin. Neurophysiol., 1976, 40: 33-42.

Skrandies, W. Data reduction of multichannel fields: global field power and principal component analysis. Brain Topography, 1989, 2: 73-80.

Skrandies, W. Scalp potential fields evoked by grating stimuli: effects of spatial frequency and orientation. Electroenceph. Clin. Neurophysiol., 1984, 58: 325-332.

Skrandies, W. Time range analysis of evoked potential fields. Brain Topography, 1989, 1: 107-116.

Skrandies, W. and Lehmann, D. Spatial principal components of multichannel maps evoked by lateral visual halfield stimuli., Electroenceph. Clin. Neurophysiol., 1982, 54: 662-667.

Small, J.G., Milstein, V., Small, I.F., Miller, M.J., Kellams, J.J. and Corsaro, C.J. Computerized EEG profiles of Haloperidol, Chlorpromazine, Clozapine and Placebo in treatment resistant Schizophrenia. Clinical EEG, 1987, 18: 124-135.

Snyder, R.D. Topographic mapping in childhood developmental dyslexia [letter], Ann Neurol, 1980, 8: 642-643.

Sparks, D.L. Neural cartography: sensory and motor maps in the superior colliculus. Brain Behav. Evol., 1988, 31: 49-56.

Spitzer, A.R., Cohen, L.G., Fabrikant, J. and Hallett, M. A method for determining optimal interelectrode spacing for cerebral topographic mapping. Electroenceph. Clin. Neurophysiol., 1989, 72: 355-361.

Spydell, J. R., Pattee, G. and Goldie, W. D. The 40 Hz auditory event-related potential: normal values and effects of lesions. Electroenceph. Clin. Neurophysiol., 1985, 62: 193-202.

Srebo, R. Localization of cortical activity associated with visual recognition in humans. JP, 1985, 360: 247-259.

Struve, F.A., Straumanis, J.J., Patrick, G. and Price, L. Topographic mapping of quantitative EEG variables in chronic heavy marihuana users: empirical findings with psychiatric patients. Clin. Electroenceph., 1989, 20: 6-23.

Sullivan, G.W., Davis, L.E., Mondt, J.P, Grace, K.M. and Flynn, E.R. Magnetoencephalographic comparison of cortical sensorimotor extrema evoked by flexion of index finger and thumb. Brain Topography, 1989, 1: 257-262.

Sutherling, W.W., Crandall, P.H., Cahan, L.D. and Barth, D.S. The magnetic field of epileptic spikes agrees with intracranial localizations in complex partial epilepsy. Neurology, 1988, 38: 778-786.

Sutton, J.P., Whitton, J.L., Topa, M. and Moldofsky, H. Evoked potential maps in learning disabled children. Electroenceph. Clin. Neurophysiol., 1986, 65: 399-404.

Taira, T., Amano, K. and Kawamura, H. et al. Significance probability mapping of brain electrical activity. Its problem and specified z-statistic mapping. Neurol. Surg., 14/3, 1986.

Taira, T. and Kitamura, K. Significance probability mapping of patients with migraine. Electroenceph. Clin. Neurophysiol., 1985, 61: 583.

Takahashi, H., Yasue, M. and Ishijima, B. Dynamic EEG topography and analysis of epileptic spikes and evoked potentials following thalamic stimulation. Appl. Neurophysiol., 1985, 48: 418-422.

Tatsuno, J., Ashida, H. and Takao, A. Objective evaluation of differences in patterns of EEG topographical maps by Mahalanobis distance. Electroenceph. Clin. Neurophysiol., 1988, 69: 287-290.

Thatcher, R.W., Krause, P.J. and Hrybyk, M. Cortico-cortical associations and EEG coherence: a two - compartmental model. Electroenceph. Clin. Neurophysiol., 1986, 64: 123-143.

Thau, K., Rappelsberger, P., Lovrek, A., Petsche, H., Simhandl, C. and Topitz, A. Effect of lithium on the EEG of healthy males and females. A probability mapping study. Neuropsychobiol., 1989, 20: 158-163.

Thickbroom, G.W., Davies, H.D., Carroll, W.M. and Mastaglia, F.L. Averaging, spatio-temporal mapping and dipole modelling of focal epileptic spikes. Electroenceph. Clin. Neurophysiol., 1986, 64/3: 274-277.

Thickbroom, G.W., Mastaglia, F.L. and Carroll, W.M. Spatio-temporal mapping of evoked cerebral activity. Electroenceph. Clin. Neurophysiol., 1984a, 59: 425-431.

Thickbroom, G.W., Mastaglia, F.L. and Carroll, W.M. Computerised topographical mapping of scalp recorded event-related potentials. Int. J. Bio-Med. Comput., 1984b, 15/2.

Thickbroom, G.W., Mastaglia, F.L., Carroll, W.M. and Davies, H.D. Source derivation: application to topographic mapping of visual evoked potentials., Electroenceph. Clin. Neurophysiol., 1984, 59: 279-285.

Thompson, C.J., Hardy, T. and Bertrand, G. A system for anatomical and functional mapping of the human thalamus. Comput. Biomed. Res., 1977, 10: 9-24.

Tomberg, C., Desmedt, J.E., Ozaki, I., Nguyen, T.H. and Chalklin, V. Mapping somatosensory evoked potentials to finger stimulation at intervals of 450 to 4000 msec and the issue of habituation when assessing early cognitive components. Electroenceph. Clin. Neurophysiol., 1989, 74: 347-358.

Tonnquist-Uhlen, I., Borg, E. and Spens, K. Auditory stimulation brain map. Scand. Audiol., 1989, 18: 3-12.

Torello, M.W. Topographic mapping of EEG and evoked potentials in psychiatry: delusions, illusions, and realities. Brain Topography, 1989, 1: 157-174.

Torello, M.W., Phillips, T., Hunter, W.W. and Csuri, C.A.   Combination imaging: magnetic resonance imaging and EEG displayed simultaneously. J. Clin. Neurophysiol., 1987, 4: 274-275.

Towle, V.L., Brigell, M. and Spire, J. Hemi-field pattern visual evoked potentials: a comparison of display and analysis techniques. Brain Topography, 1989, 1: 263-270.

Tranel, D., Damasio, A.R. and Damasio, H. Intact recognition of facial expression, gender, and age in   patients with impaired recognition of face identity. Neurology, 1988, 38: 690-696.

Tsuji, S. and Murai, Y. Scalp topography and distribution of cortical somatosensory evoked potentials to median nerve stimulation. Electroenceph. Clin. Neurophysiol., 1986, 65: 429-439.

Tucker, D.M. and Roth Bair, T.B. Functional connections among cortical regions: topography of EEG coherence. Electroenceph. Clin. Neurophysiol., 1986, 63: 242-250.

Tukey, J.W.   "Exploratory data analysis" Addison-Wesley, Reading, Mass., 1977.

Udin, S.B. and Fawcett, J.W. Formation of topographic maps. Annu. Rev. Neurosci., 1988, 11: 289-327.

Ueno, S., Hasegawa, R. and Harada, K. Automatic interpretation of the brain wave. Iyodenshi To Seitai Kogaku, 1976: 14(5): 379-386.

Ueno, S. and Matsuoka, S. Topographic computer display of abnormal EEG activities in patients with brain lesions. Digest of the 11th Int. Conf. on Med. and Biol. Eng., Ottawa, 1976a, 218-219.

Ueno, S. and Matsuoka, S. Extraction of abnormal EEG with slow waves and its display method. Iyodensi To Seitai Kogaku, 1976b, 14(2): 118-124.

Ueno, S., Matsuoka, S., Mizoguchi, T., Nagashima, M. and Cheng, C.L.

Topographic computer display of abnormal EEG activities in patients with CNS diseases. Memoirs of the Faculty of Engineering. Kyushu University, 1975: 34(3), 195-209.

Vantoller, S. and Reed, M.K. Brain electrical-activity topographical maps produced in response to olfactory and chemosensory stimulation. Psychiatry Research. 1989, 29: 429-430.

Vaughan, H.G., Jr, Topographic analysis of brain electrical activity. Electroenceph. Clin. Neurophysiol. 1987, Supp. 39: 137-142.

Vaughan, H.G., Jr, Costa, L.D. and Ritter, W. Topography of the human motor potential. Electroenceph. Clin. Neurophysiol., 1968, 25: 1-10.

Vaughan, H.G., Jr and Ritter, W. The sources of auditory evoked responses recorded from the human scalp. Electroenceph. Clin. Neurophysiol., 1970, 28: 360-367.

Verma, N.P. and Delacruz, C.R. Brain electrical-activity mapping of monocular P100 reveals another type of uncrossed asymmetry. Clin. Electroenceph., 1989, 20: 254-258.

Verma, N.P., Nichols, C.D., Greiffenstein, M.F., Singh, R. P. and Hurst-Gordon, D. Waves earlier than P3 are more informative in putative subcortical dementias: a study with mapping and neuropsychological techniques. Brain Topography, 1989, 1: 183-192.

Vieth, J., Schueler, P., Harsdorf, S. V., Fischer, H. and Grimm, U. AC-MEG and AC-EEG at verified focal lesions and DC-MEG shifts during seizure and interictal periods. Paper presented at the Annual Meeting of the American Epilepsy Society, San Francisco, California, October, 1988.

Vollmer, R., Petsche, H., Pockberger, H., Prohaska, O. and Rappelsberger, P. Spatiotemporal analysis of cortical seizure activities in a homogeneous cytoarchitectonic region. In: M.A.B. Brazier and H. Petsche (Eds.), Architectonics of the Cerebral Cortex. Raven Press, New York, 1978, 281-306.

Vrba, J,. Fife, M., Burbank, M., Weinberg, H. and Brickett, P. Spatial discrimination in SQUID gradiometers and 3rd order gradiometer performance. Can. J. Phy., 1982, 60: 1060-1073.

Vvedensky, V.L., Ilmoniemi, R. J. and Kajola, M. J. Study of the alpha rhythm with a 4 channel squid magnetometer. Med. and Biol. Engineering and Computing, 1985, Supp. 23 (Part 1): 11-12.

Wallin, G. and Stalberg, E. Source derivation in clinical routing EEG. Electroenceph. Clin. Neurophysiol., 1980, 50: 282-292.

Walter, W. G. The living brain. Duckworth, London, 1953.

Walter, D.O. and Brazier, M.A.B. Advances in EEG analysis. Electroenceph. Clin. Neurophysiol., 1969, Supp. 27, 1-78.

Walter, D.O., Etevenon, P., Pidoux, B., Tortrat, D. and Guillou, S. Computerized topo-EEG spectral maps: difficulties and perspectives, Neuropsychobiol., 1984, 11: 264-272.

Walter, D.O., Kado, R.T., Rhodes, J.M., and Adey, W.R. Electroencephalographic baselines in astronaut candidates estimated by computation and pattern recognition techniques. Aerospace Med., 1967, 38: 371-379.

Walter, D.O., Rhodes, J.M., Brown, D. and Adey, W.R. Comprehensive spectral analysis of human EEG generators in posterior cerebral regions. Electroenceph. Clin. Neurophysiol., 1966, 20: 224-237.

Walter, W.G. and Shipton, H.W. A new toposcopic display system. Electroenceph. Clin. Neurophysiol., 1951, 3: 281-292.

Weinberg, H., Brickett, P. Gordon, R. and Harrop, R. The interaction of thalamo-cortical systems in the 40 Hz following response Advisory Group for Aerospace Research and Development, 1987, 432, 12: 1-6.

Weinberg, H., Brickett, P., Coolsma, F. and Baff, M. Magnetic localisation of intracranial dipoles: simulation with a physical model. Electroenceph. Clin. Neurophysiol., 1986, 64: 159-170.

Weinberg, H., Brickett, P. A., Deecke, L. and Boschert, J. Slow magnetic fields preceding movement and speech. Il Nuovo Cimento, 1983, 2D 1: 495-504.

Weinberg, H., Brickett, P., Neill, R. A., Fenelon B. and Baff, M. Magnetic fields evoked by random-dot stereograms. In: H. Weinberg, G. Stroink and T. Katila (Eds.), Biomagnetism: Applications and Theory, Pergamon Press, New York, 1985, 354-359.

Weinberg, H., Brickett, P., Robertson, A., Crisp, D., Cheyne, D. and Harrop, R. A Study of generators in the human brain associated with stereopsis. Paper presented at the Advanced Group for Aerospace Research and Development (NATO) Conference, Trondheim, Norway, 1987.

Weinberg, H., Brickett, P., Robertson, A., Harrop, R., Cheyne, D. O., Crisp, D., Baff, M. and Dykstra, C. The magnetoencephalographic localisation of generator-systems in the brain: early and late components of event related potentials. J. Alcohol, 1987, 4: 339-345

Weinberg, H., Brickett, P. A., Vrba, J., Fife, A. A. and Burbank, M. B. The use of a SQUID third order spatial gradiometer to measure magnetic fields of the brain. In: R. Karrer, J. Cohen and P. Teuting (Eds.), Brain and information: event related potentials. New York Acad. Sci., 1984, 42: 743-752.

Weinberg, H., Cheyne, D., Brickett, P., Gordon, R. and Harrop, R. The interaction of thalamocortical systems in the 40 Hz following response. Paper presented at the Advanced Group for Aerospace Research and Development (NATO) Conference, Trondheim, Norway, 1987.

Weinberg, H., Crisp, D., Brickett, P., Harrop, R., Purves, S. J., Li, D.K.B., Jones, M. W. and Baff, M. The combination of MEG and MRI in the estimation of generators associated with interictal discharges. In: S. N. Erne and G. L. Romani (Eds.), Functional localization: a challenge for biomagnetism. World Scientific: Singapore, 1987.

Weinberg, H., Johnson, B., Cohen, P., Crisp, D. and Robertson, A. Functional imaging of brain responses to repetitive sensory stimulation: sources estimated from EEG and SPECT. Brain Topography, 1989, 2: 171- 180.

Weinberg, H., Robertson, A., Brickett, P., Cheyne, D., Harrop, R., Dykstra, C. and Baff, M. Functional localization of current sources in the human brain associated with the discrimination of moving visual stimuli. In: R. Johnson Jr. R. Parasuraman and J.W. Rohrbaugh (Eds.), Current trends in event-related potential research. Electroenceph. Clin. Neurophysiol., (Suppl.40): 1987.

Weinberg, H., G. Stroink and T. Katila, (Eds.), Biomagnetism: Applications and Theory. Pergamon Press, New York, 1985.

Welker, W., Mapping the brain. Historical trends in functional localization. Brain Behav. Evol., 1976, 13: 327-243.

Williamson, S.J., Romani, G. L., Kaufman, L. and Modena, I. (Eds.), Biomagnetism: an interdisciplinary approach. Plenum Press, New York, 1983, Nato ASI Series A: 66.

Williamson, S.J. and Kaufman. L.   Advances in neuromagnetic instrumentation and studies of spontaneous brain activity. Brain Topography, 2: 129-139.

Wong, P.K.H. Stability of source estimates in rolandic spikes. Brain Topography, 1989, 2: 31-36.

Wong, P.K.H. Topographic EEG analysis.   In: J. Wada and R. Ellingson, (Eds.), Handbook of EEG and Clinical Neurophysiology, Revised Series. Epilepsy. Elsevier, 1990, Vol. 4, Chp. 15 (in press).

Wong, P.K.H., Bencivenga, R. and Gregory, D. Statistical classification of spikes in benign rolandic epilepsy. Brain Topography, 1989, 1: 123-130.

Wong, P.K.H., Bencivenga, R., Jan, J.E. and Farrell, K. Detection of visual field defect using topographic evoked potential in children. In: G.C. Woo (Ed.), Low vision: principles and applications. Springer-Verlag, New York, 1987, 168-179.

Wong, P.K.H., Farrell, K., Jan, J.E. and Whiting, S. Preliminary study of topographic visual evoked potential mapping in children with permanent cortical visual impairment. In: G.C. Woo (Ed.), Low Vision: Principles and Applications. Springer-Verlag, New York, 1987, pp. 180-189.

Wong, P.K.H. and Gregory, D. Dipole fields in rolandic discharges. Amer. J. EEG Technol., 1988, 28: 243-250.

Wong, P.K.H., Gregory, D. and Farrell, K. Comparison of spike topography in typical and atypical benign rolandic epilepsy of childhood. Electroenceph. Clin. Neurophysiol., 1985, 61: S47.

Wong, P.K.H., Jan, J.E. and Farrell, K. Topographic VEP mapping in cortical visual impairment. Electroenceph. Clin. Neurophysiol., 1985a, 61: 523.

Wong, P.K.H., Jan, J.E. and Farrell, K. Topographic VEP as an aid in the diagnosis of visual field defects in young children. Electroenceph. Clin. Neurophysiol., 1985b, 61: 599.

Wong, P.K.H., Bencivenga, R., Gregory, D. Statistical classification of spikes in benign rolandic epilepsy. Brain Topography 1988, 1: 123-129.

Wong, P.K.H. and Weinberg, H. Source estimation of scalp EEG focus. In: G. Pfurtscheller and F. Lopes da Silva (Eds.), Functional Brain Imaging. Hans Huber Publishers, Toronto, 1988: 89-95.

Wood, C.C., Spencer, D.D., Allison, T., McCarthy, G., Williamson, P.D. and Goff, W.R. Localization of human sensorimotor cortex during surgery by cortical surface recording of somatosensory evoked potentials. J. Neurosurg., 1988, 68: 99-111.

Wood, F. Focal and diffuse memory activation assessed by localized indicators of CNS metabolism: the semantic-episodic memory distinction. Hum. Neurobiol., 1987, 6: 141-51.

Woolsey, C.N., Erickson, T.C. and Gilson, W.E. Localization in somatic sensory and motor areas of human cerebral cortex as determined by direct recording of evoked potentials and electrical stimulation. J. Neurosurg., 1979, 51: 476-506.

Yamada, T., Graff-Radford, N.R., Kimura, J., Dickens, Q.S. and Adams, H.P. Jr. Topographic analysis of somatosensory evoked potentials in patients with well-localized thalamic infarctions. J. Neurol. Sci., 1985, 68: 31-46.

Yamada, T., Kayamori, R., Kimura, J. and Beck, D.O. Topography of somatosensory evoked potentials after stimulation of the median nerve. Electroenceph. Clin. Neurophysiol., 1984, 59: 29-43.

Yingling, C.D. and Hosobuchi, Y. A subcortical correlate of P300 in man. Electroenceph. Clin. Neurophysiol, 1984, 59: 72-76.

Young, I.R., Khenia, S., Thomas, D.G., Davis, C.H., Gadian, D.G., Cox, I.J., Ross, B.D. and Bydder, G.M. Clinical magnetic susceptibility mapping of the brain. J. Comput. Assist. Tomogr. Jan-Feb 1, 11: 2-6.

Zeitlhofer, J., Saletu, B., Anderer, P., Asenbaum, S., Spiss, C., Mohl, W., Kasall, H. and Wolner, E. Topographic brain mapping of EEG before and after open-heart surgery. Neuropshychobiol., 1988, 20: 51-56.

Zhang, J.Z., Li, J.Z. and He, Q.N. Statistical brain topographic mapping analysis for EEGs recorded during Qi Gong State. International J. Neurosci., 1988, 38: 415-425.

Zulch, K.J. A critical appraisal of "Lokalisationslehre" in the brain., Naturwissenschaften, 1976, 63: 255-265.

GLOSSARY

ADC: analog to digital conversion; converts analog data into computer-ready digital format.

Alias: an ADC error which is due to too low a sampling rate, and which may result in abberent or ghost output components. This error is unpredictable and impossible to detect or correct for.

Ambiguous solution: or non-unique solution, refers to dipole localization method, where there are many possible and mathematically equal inverse solutions.

ANOVA: analysis of variance, a form of generalized t-test, frequently used to test for equality of the means among different groups.

BP (Bereitschaft Potential): a cerebral "readiness potential" produced under specific cognitive conditions.

BREC (Benign Rolandic Epilepsy of Childhood): a common form of benign focal epilepsy seen in children from age 3 to 14 years. There is predominantly oral-facial involvement during seizures, normal neurological examination, intellectual functioning, and absence of brain lesion.

CART (Classification and Regression Trees): a non-parametric classification technique using a process of systematic partitions. It is used in exploratory data analysis.

Cartography: the mechanical aspects of displaying 3 dimensional data (e.g., scalp EEG) onto a 2 dimensional flat surface.

CNV (Contingent Negative Variation): a slow negative polarity brain potential elicited by the expectation of a stimulus.

Confirmatory analysis: a specific approach in contrast to exploratory analysis - a hypothesis had been set up to be tested, within the context of a properly designed experiment where a set of data had been collected.

Constraints: refers to limits placed on the parameters for a source modelling procedure, an example being the constraint that source locations must be intracranial.

Correlation: the degree to which one variable is associated with another. The Pearson's coefficient (r^2) is often used to measure correlation between 2 variables that are linearly related.

Dewar: a super-insulating container used to keep SQUID at ultra-low temperatures. Usually contains circulating liquid helium as the coolant.

d.f. (degrees of freedom): a mathematical parameter which may be taken to represent the dimensionality or complexity of a system.

Digital filter: frequency filtering performed on the digital data by computer software, which can mimic the high-pass, low-pass, band-pass and band-stop (notch) characteristics of analog filters. If designed properly, can achieve very sharp roll-off slopes.

Dipole: can mean either a hypothetical current generator (current dipole or source), or an electric field pattern with simultaneous positive-negative polarities (dipole topography). This ambiguity must be resolved within the context of the term.

Dipole localization (or dipole localization method, DLM): the method of mathematically estimating sources from the scalp topography of EEG or MEG data.

Distributed system: usually refers to a neuronal generator configuration that is not representable by a discrete location ("point source"). This may be an extensive dipole sheet as found on a flat area of cortex.

Dynamical analysis: a method of analysis of the behaviour of a system which evolved from non-linear control theory and differential calculus; it explores the number of deterministic factors (dimensions) that can characterize such a system. Includes the study of fractals, chaos, strange attractors, Lyapanov exponents.

EP (Evoked Potential): cerebral electrical responses to sensory stimuli (auditory, visual, somatosensory, etc.).

Exploratory analysis: a general approach to data analysis such that many different approaches and tests are tried, in order to generate insight into possible trends and associations. However, no firm conclusion can be made as no specific hypothesis had been posed for testing.

Factor analysis: similar to principal component analysis (PCA), except that the number of factors are fixed, then the data examined for the linear combination of these factors which can best account for the data variability.

Forward solution: assuming that the material constants relating to electrical conductivities of the skull, scalp, CSF, parenchyma etc. are known, the forward solution is the calculation of the scalp electric field based on a set of known source parameters. This can be applied to MEG data in similar fashion.

Fourier transform or Fast Fourier Transform (FFT): breaks down a time-series signal (EEG or MEG) mathematically into constituent frequency components, e.g., delta, theta, alpha, beta.

FT (Femtotessla): a commonly used unit of magnetic field strength.

Gaussian: the statistical characteristic of being normally distributed, i.e., having a "bell-shaped" distribution, implying that a particular mathematical formula may be used for definition.

Global Field Power (GFP): a measure of the "hilliness" of a topographic distribution or map.

Gradient: the steepness of the electric (or magnetic) field potential changes at a given point of a potential map.

Gradiometer: a device sensitive to minute changes in the strengths of magnetic flux, similar to differential amplifiers in EEG.

h^2 a non-linear correlation value, calculated by use of the non-linear regression algorithm developed by Lopes da Silva and Pijn.

Head model: a mathematical description of the characteristics of the human head, including the shape of the skull, presence of holes, conductivities of the media (skin, muscle, bone, CSF, parenchyma) etc. It may be as simple as a solid sphere (where all the above different media are lumped together, called a single-shell or single-sphere model), to a 3-shell model (separate shells of bone, CSF and brain parenchyma), and finally to the so-called realistic head models (where the head is depicted with anatomic and physical accuracy).

Hjorth derivation or source derivation: a reference transformation using mathematical calculation whereby each electrode reflects only locally generated signals, and far-field (common) signals are attenuated. The advantages are: enhancement of focal signals, attenuation of diffuse signals, relatively free of reference contamination.

Horizontal dipole or tangential dipole: a source orientated 90° to the radial direction, i.e. pointing towards the horizontal, or tangentially to the skull.

Isocontour: contour lines in a potential map of the same strength.

Interpolation: mathematical calculation of unknown values lying between 2 or more known values; generally used to estimate what the scalp voltage may be in a region where there is no actual recording electrode present. Common schemes include nearest 3 or 4 neighbours (linear), quadratic, cubic, or spline (non-linear) algorithms.

Integral: the area under the curve of a given tracing.

Inverse solution: the reverse of the forward solution, that is the estimation of the source parameters based on a known scalp electric field. Also called dipole localization method or least-squares dipole fit, referring to the mathematical approximation procedures used.

Laplacian: the spatial derivative of a potential field, or the gradient indicating how quickly the field changes over a given distance. Similar to estimates of current source density. It is usually implemented by the Hjorth derivation method.

Magnetometer: a device for recording magnetic field strength.

Mahalanobis distance: a measure of the standardized distance separating 2 clusters of variables.

Maxima: the point of greatest positive voltage in a given topography.

MEF (Magnetic Evoked Fields): analogous to EPs for EEG.

MEG (Magnetoencephalography): measurement of the magnetic signals from the brain based on extracranial recording devices.

Minima: the point of greatest negative voltage in a given topography.

MP (Motor Potential): the MEG equivalent of the Bereitschaft Potential (BP).

MRI (Magnetic Resonance imaging): or nuclear magnetic resonance imaging.

Multivariate: multiple variables are considered together in a statistical analysis; as opposed to univariate.

Nyquist frequency: the minimum ADC rate that can correctly sample a given input data. Theoretically this is double the maximum frequency contained in the input.

One-dipole model: a single source is assumed to give rise to a given scalp field, so the estimation of the inverse solution will yield 1 set of source parameters.

P300: a particular auditory evoked potential; the P300 is the positive peak elicited by rare or unexpected stimuli at approximately 300 msec. latency.

PCA (Principal Component Analysis): a method of breaking down the input data into factors or components to account for as much of the variability of the data as possible.

Peak (in the spatial sense): refers to a maxima or minima.

Precision: the number of bits of ADC precision determines the accuracy of the converted result. Usually 8 bits are used, but 10 or 12 bits are superior.

Post-processing: analysis or computation on the acquired data, usually performed by computer software.

Propagation: neuronal conduction of a signal.

r^2: the Pearson linear correlation coefficient.

Radial dipole: a source orientated along the direction of the radius of the skull (usually taken to be locally spherical), giving rise to a scalp field that has a single polarity.

Reformatting: rearrangement of multi-channel data into a montage other than as recorded, usually for optimal display reasons. Requires that the original data be in a referential form to start with.

SI (Stability Index): a relative mathematical measure reflecting the degree of stability of an inverse solution in a given situation. Described by Wong.

Spatial density: the concentration of recording electrodes (or points) used to construct the topographic distribution of scalp signals.

Spatial-temporal: consideration of both location and chronologic features; this usually means taking into account both inter-channel differences and also behaviour over time, and applies to how inverse solution is performed.

SQUID (Superconducting Quantum Interference Device): a device that can measure minute magnetic signals, based on the principles of low-temperature physics.

Source: a mathematical description of a neuronal generator, usually involving 6 parameters to define location, strength, configuration and direction.

Source derivation: see Hjorth derivation.

Source parameters: a set for description of location, and another set to describe direction and strength of source. Both may be expressed in cartesian coordinates: location as (x,y,z), while the direction and strength may be denoted by (x',y',z').

Spatial Filter: a filter that alters the spatial characteristics of a topographic dataset. The Laplacian is one such example.

Spike apex: the peak negativity of a spike, at the channel of greatest amplitude.

Spike map: the topographic analysis of spike discharges, perhaps aided by averaging many spikes from the same focus.

SPM (Statistical Probability Mapping): a limited method of mapping the statistical z or t value of a given map.

Spread: the apparent movement (or translocation) of a peak across a topography over time.

SSD (Sum Squared Deviation): an error term involved in DLM.

Tangential dipole: see horizontal dipole.

Time-series: data that varies over time, like EEG, MEG or EP. Such data can be subjected to statistical treatment based on traditional techniques.

Topography (EEG): the spatial pattern of electrical field, usually recorded from the scalp from multiple electrodes. Commonly 19 channels are used, although up to 128 have been reported.

Two-dipole model: same as a one-dipole model, except that 2 sources are assumed in the calculations.

Univariate: a single variable is considered at a time.

Volume conduction: the instantaneous spread of a signal across the scalp; has definite spatial characteristics.

INDEX